Wildlife Adventures
In the Canadian West

DICK DEKKER

Rocky
Mountain Books

Cover photos. Photos of author: Brian Genereux.
Photo of wolf: Monty Sloan/wolfphotography.com.
Photos of elk: Martin de Jonge.

Editing, line drawings, and photographs by the author unless otherwise indicated.

 Published by Rocky Mountain Books
#4 Spruce Centre SW, Calgary, AB T3C 3B3
Printed and bound in Canada by
RMB Houghton Boston, Saskatoon

We acknowledge the financial support of the Government of Canada through the Book Publishing Industry Development Program (BPIDP) for our publishing activities.

National Library of Canada Cataloguing in Publication Data

Dekker, Dick, 1933–
  Wildlife adventures in the Canadian west

  Includes bibliographical references.
  ISBN 1–894765–36–2

  1. Dekker, Dick, 1933– 2. Zoology—Canada, Western—Anecdotes. 3. Outdoor life—Canada, Western—Anecdotes. 4. Natural history—Canada, Western—Anecdotes. 5. Canada, Western—Description and travel. 6. Yukon Territory—Description and travel. 7. Zoologists—Canada—Biography. 8. Adventure and adventurers—Canada, Western—Biography. I. Title.
QL221.W47D44 2002   591.9712   C2002–910971–X

TABLE OF CONTENTS

Introduction

Quest For Wildness

The cartoon illustration pictured an isolated, snow–bound cabin besieged by a pack of howling wolves, in the mountains of Yukon. Although it is more than half a century ago that I saw that drawing, I remember it to this day. The imagery touched a chord in me, not by its contrived terror, but because of the notion of virgin northern wilds. Why a kid living in one of the most densely populated countries in the world should feel an affinity for such a stark, sub–arctic scene is a question of wonder. Had it conjured up echoes of an earlier life, or was it a vision for the future?

A few years later, I was similarly touched, but in a negative context, when the international news media included a story about wolves in Alaska. The state had instituted a bounty, and most upsetting of all, local bush pilots considered it a profitable sport to shoot the predators from the air.

I was deeply shocked that such callous, government sponsored killing was taking place in a corner of the world which I had come to view as an untrammelled refuge for nature. Today, respect for wildlife is a sentiment shared by millions of North Americans as well as Europeans. Wilderness has become a cherished concept and the wolf its celebrated ambassador. The world over, Canadians are envied for having both these treasured assets from sea to shining sea.

In 1959, in search of the wild, I moved to Alberta where I found scenic beauty and fascinating animals in abundance. However, the pristine, untouched wilderness was as elusive as ever. At that time, the wolf had been exterminated in all of the great National Parks in Alberta's Rocky Mountains. As a consequence, the presence of this predator, which is no more and no less than a part of the whole ecosystem, became the standard by which I measured the wild.

My visceral passion for a northern landscape where the animal and plant communities were complete and unspoilt by mankind led me farther afield, until my greenhorn dreams almost cost me my life. In those early years I paddled hundreds of miles down swift, Yukon rivers. Later, Irma and I canoed the Peace River in northwestern Alberta and several remote lake systems in northern Saskatchewan.

The long trail of exploration eventually veered back to the Rocky Mountains in Alberta. In the meantime, public and government attitudes about wildlife and its management were going through a metamorphosis, and today much of the Rockies are preserved as a paradise–like sanctuary for animals which I had sought far and wide.

Over four decades, I have intensely observed the intertwined lives of prey and predators in Jasper, jewel in the crown of Canadian National Parks. My interest was driven solely by a deep, personal sense of curiosity, and my primary concern was not to cause any disturbance. I have done my studies in my own time and at my own cost. However, from time to time, to help defray travel expenses I received a small grant from World Wildlife Fund Canada, the Canadian Wolf Defenders, and the Alberta Recreation, Parks and Wildlife Foundation. For information and logistical support in the field, I am indebted to Parks Canada and the individual wardens named in these chapters. On my part, I have made a modest contribution to the literature on the park's fauna by the publications listed in the appendix.

My 1994 book Wolf Story (and its 1998 update Wolves of the Rocky Mountains from Jasper to Yellowstone) largely concentrated on the published reports of researchers who studied the history, management, and science of wolves. A few years ago, one of the reviewers of Wolf Story concluded that its epilogue "will make readers wish that Dekker would write a book about his adventures and misadventures during his thirty years in the Canadian wilderness he loves and defends."

This is that book, my own story, describing my travels and wildlife observations in all seasons of the year. From its beginnings, the narrative flows like a clear, untouched wilderness stream, with lively riffles and quiet pools that give pause for reflection and perspective. Going with the flow, I hope that the reader experiences the same thrill of suspense and excitement that has spurred me on along my journey of discovery, a journey that eventually led to the realization of a boyhood dream, a snow–laden cabin serenaded by a pack of wolves under a cold, star–studded sky.

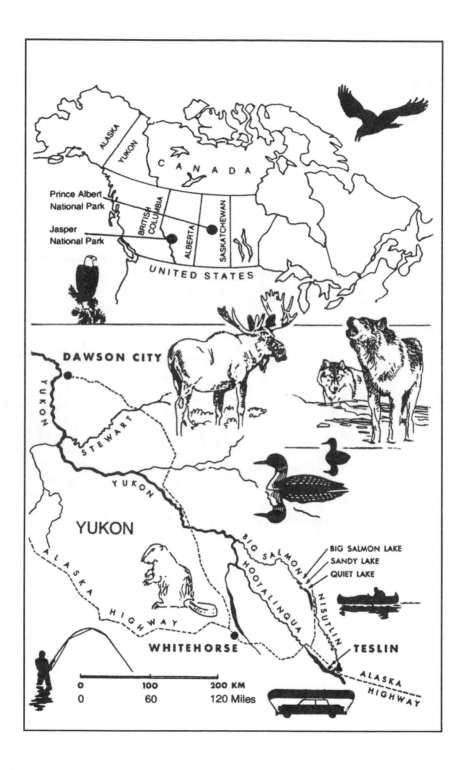

Part I

Greenhorns In The Wilderness

Chapter 1

Down the River of Lots

"We'll see lots!" said George, my young Indian guide, as we began our journey down the Nisutlin River in Yukon.

"Any bears and wolves?" I asked.

"Lots!"

"What about lynx and otter?"

"Lots!"

His positive predictions were music to my ears, reinforcing my own high expectations. I had wanted to make a wilderness trip like this for a very long time, and we were finally on our way. Sitting in the bow of the big freighter canoe I glanced back at George with a grin of satisfaction. Stowed aboard between us was everything we needed for the next several days: camping gear, warm clothes, an axe, cooking pots, and boxes of groceries. On top of the load, within easy reach of George, was his high–powered rifle and a small calibre "twenty–two." As my hired guide he was to steer the canoe and, by his own suggestion, provide us with wild meat. "I'll get lots," he had said with a self–assured expression on his face. "Ducks, geese, rabbits…"

"Lots" had been his standard reply to all of my inquiries. He was apparently a born optimist. Perhaps that was the best attitude to have in the wilderness. To be negative or insecure, like some city people who have lost touch with nature, could have lethal consequences here. A child–like trust in the Great Provider was the only defence against the harsh realities of life in the Canadian North. To feel at home and survive in this lonely land, one had to have a solid core and be at peace with one's place in the universe. Such spiritual fortitude is bred into the bone of aboriginals by nature's predictability. The sun rises in the morning and sets at night. These everyday events, banal as they may seem to those who live in the city, are still the greatest spectacles on earth. In town you never knew when the power went out, the tube quit, or the car broke down. In nature you could set your compass by a few simple truths, such as that water flows downhill.

These thoughts occurred to me very much later, after I had spent a great deal of time in the wilds myself, often alone, and after my obsession to see unspoilt wilderness had been tempered by reality.

When the Yukon adventure started, in 1961, I had the same youthful confidence as George, although our perspectives and background were quite different. He had made a living here all his life, and I was a naive, city–bred idealist who had come from halfway across the globe. An immigrant to this country, a Canadian by choice, my main interest lay in the wildlife of this vast land. Romanticizing its virgin northwoods as the Holy Grail of landscapes, I had read several books written by trappers who blazed their trail to the Yukon during the 1930s. At that time, while the economy of the industrialized world had collapsed, the skins of furbearing animals were worth more than their weight in gold. A silver fox could fetch 500 dollars, a small fortune in those days. However, the lure that had drawn me irresistibly north had nothing to do with money. All I wanted was to see the boreal wilderness and its indigenous mammals and birds.

I had met George quite by chance soon after arriving in the Yukon. The old Pontiac loaded up with my few earthly possessions, I had driven north from Alberta on the renowned Alaska Highway. The road had been built for military purposes by the U.S. Army during the second World War. The Canadian portion of the route, starting in Dawson Creek, British Columbia, was some 1500 miles (2500 km) long, bulldozed across untouched country. Along the way, it connected a few isolated, mostly native hamlets. The surface is paved now, but in 1961 nearly all of it was gravel which alternately turned to mud or dust, depending on the weather. Fortunately, in those days traffic was light.

Penetrating deep into the northern forest, the narrow road passed by endless miles of mountains, valleys, lakes, and rivers, but it gave me little satisfaction just to look at all this pristine beauty through the windshield. What I wanted was to get a feel of the country far away from roads and other signs of human occupation.

By the time I had reached the Indian village of Teslin in the Yukon Territory, the urge to stop and leave the highway behind had grown too strong to resist. On both sides of town glittered large lakes. Beyond were gently sloping mountains, their lower flanks clothed in forests unmarred by clearcuts.

Striking up a conversation with the owner of the only gas station, I asked for advice: "Would any of the rivers around here be safe for a small canoe?"

"Oh, just about all of them," the man said: "In these wide valleys, the fall in elevation is very small. You can take the Nisutlin, the Hootalinqua, the Stewart, the Yukon…"

Pointing them out on a large map displayed on the wall of his general store, he casually mentioned that the local canoe expert was a priest, an Irish lad. Perhaps I should go and see him.

One glance at the small, twelve–foot (3.6 m) canoe I had carried along on my car, convinced the young priest that it was not quite suitable for any trip. How about if he lent me his big nineteen–foot (5.7 m) freighter? And a good river to start with was the Nisutlin, right here by Teslin. All I needed was someone to drive me to a convenient starting point some hundred miles (160 km) north of here. Then I could simply float back to town.

Great idea! But how would I, a greenhorn by any standard, ever be able to handle that heavy canoe?

"Why don't you ask a local boy to come with you?" the priest suggested. "George is a good man and he knows the river. You pay him five bucks a day, and I'll charge you the same for the rental of my canoe."

The house of George's parents was small and dilapidated, paint peeling on the outside, wall paper in shreds inside. But the family seemed happy and completely at leisure. All they did during summer was go fishing. George was the only son. As he told me later, his five older brothers had died during the 1940s of diseases introduced into this isolated wilderness by the road construction crews.

George was indeed quite willing to come along on the river trip and we arranged to leave the very next day. As to someone who could drive my car back from the starting point, he would ask his cousin Frank who had a driver's licence. In return for his services, he would surely be content with a few cans of food.

That night I camped on the edge of town, to be awakened next morning by the croak of a raven. This was a good omen, I reckoned. Companion to the wolf, the raven was another legendary representative of the wilderness I had come to find. By the time I got to town, Frank and George were ready to go and all we needed to do now was to get their favourite dog, named Cisko.

"Good to have a dog along," said George: "He'll warn us of bears at night."

In a clearing of the bush behind the house, we were greeted with a commotion that made all further conversation impossible. Chained to stakes and trees, more than a dozen large Huskies were going berserk with excitement. These were the hardy beasts of burden that pulled the family's toboggan over the winter trails on the trap line. By the time of my visit, the dogs' modern replacement, the snowmobile, had not been invented yet. The Huskies remained chained up all summer and were fed exclusively with fish netted in the lake. After freeing Cisko, we were glad to get away from the uproar of envious snarls, whines, and howls.

The three of us and the dog barely fit in the car which also contained camping gear and food. The large canoe was tied onto the roof rack. Leaving the Alaska highway, we turned off on the Canol Road to Norman Wells, a sub–arctic oil facility also developed during the Second World War. After several hours of bumpy driving, we arrived at a point where the road ran parallel to the Nisutlin valley.

"Where's the river?" I wondered aloud, when I was told to stop the car.

Walking a few steps into the sloping forest, my companions pointed down. Just visible between the trees, skirting the base of the steep hill, shimmered an oxbow of water wrinkled with turbulence. Far down the valley, we could glimpse other stretches of river snaking its way through trackless forests. It was an awesome sight that gave me a queasy feeling in the stomach.

My nervousness was quickly forgotten as we set to work. It was a tricky job to get the heavy canoe down the hill. Hanging onto the trees, we carefully lowered our fragile craft into the water alongside the muddy bank. With three guys carrying the gear down, loading was quickly accomplished, and soon we said goodbye to Frank, who would drive my car back to town, visiting some relatives along the way.

~

What a relief to finally push off! George, sitting in the stern, took care of the steering and I occasionally assisted with paddling in front. The current did all the work, dragging us along at better than six miles (10 km) per hour, occasionally speeding up in riffles and rapids. Less than fifty yards (55 m) wide, the Nisutlin meandered erratically through the forest, each bend opening up new vistas. I frequently raised the binoculars and scrutinized the banks ahead, but other than trees and more trees, there was little to be seen. The river was high. It was June, and the mountain snows in the upper drainage were just beginning their big melt–down. The swollen water was milky brown with suspended silt. Later in summer, as George assured me, it would clear and the level drop. Then, open points and sand bars, now inundated, would fall dry, increasing one's chances of seeing wildlife.

All we saw that first day were a few birds, including several mergansers. "Fish ducks, no good to eat," remarked George.

As we floated by a tree supporting a huge stick nest, a bald eagle flew off and called in protest. Its head and tail showed brilliantly white against the dark background.

"I could have shot that one easily," said George presently.

"Why would you want to do that?" I asked.

"They kill beavers, young beavers. I have seen it myself."

So what, I thought, but why argue? In earlier conversations, George had already made it clear that he was not as enamoured of predatory animals as I. Wolves were no favourite of his. There were "lots" of them around. Once he counted the tracks of a pack of seventy–five! Wolves were cruel too, and they killed "lots" of moose. His uncle had come across a cow that had been partly eaten but was still alive. This kind of thing seemed hard to believe to me then. However, I now know that wolves may indeed wound large prey and leave them alone, returning later after the animal has succumbed to its wounds.

Wolf tracks, deeply imprinted in soft mud, were the first thing I saw when we at last stopped for the day. Our general direction had been south, with the hot afternoon sun reflecting off the water. Bleary–eyed as well as stiff and tired, we had decided on an early camp. George had been looking for a suitable landing site for some time, until he suddenly steered the canoe into shore. Here, there was a narrow strip of mud below a four foot (1.2 m) high bank. Stepping out and sinking ankle–deep in the ooze, we secured the boat to a tree and clambered up. The virgin forest on shore was carpeted with thick moss, very comfortable as a base for a small tent, but try to find a level spot large enough!

The first thing George did was make a fire. And not just any fire! The flames shot up high into the air under the tinder dry trees, and still he was adding armfuls of dead branches. "Lots of mosquitoes!" he grinned by way of explanation.

There were indeed lots and lots of bugs! George's idea was to burn them off. On the river, we had been practically free of this boreal plague, but in the shelter of the woods the air was thick with them. Insect repellent helped a little to reduce our misery, and soon we had organized our camp well enough so that we could start thinking about food.

It was a pleasure to watch George's skill at installing what he called a tea–stick. With one blow of his well–honed axe, he cut a poplar sapling, removed the branches and shortened it to about five feet (1.5 m). Then he sharpened one end, and, with his knife, cut a few notches in the other. After jamming the butt into the soft ground a few feet away from the fire, the tip of the tea stick protruded over the flames. In one of the notches, he hung the billy can, a fire–blackened metal pot filled with water. It boiled in minutes.

Since my boy scout days, I had camped for many years but the problem of cooking over an open fire had often got the better of me. It was heartbreaking to see the stones or sticks that supported the pot collapse before the contents boiled, not only leading to the loss of my soup or coffee but also to the dousing of the fire. George's system, a common practice all over the Canadian northwoods, was a valuable lesson learned. His tea stick even allowed one to regulate the heat, as it were, by simply raising or lowering the tip.

The food I had brought for both of us was very basic: cans of beans, spaghetti, vegetables, and fruit, Dutch cheese, bread, and cookies. George was not impressed. "I want meat!" he said: "Tomorrow I'll shoot us some meat."

At this far northern latitude and time of year, the sun barely sets below the horizon and it stayed light for twenty–four hours. It was hard to fall asleep. All night long a thrush sang the same high–pitched, monotonous phrase. High over the forest, snipes and night hawks bleated and swooshed. Their odd mechanical love calls were the result of air vibrating over their feathers as the birds rushed steeply down.

Early next morning, I awoke with a start by a loud yelping. Cisko! Had he been grabbed by a bear or a wolf? Not so! His enemies were much smaller but very, very much more numerous: mosquitoes! Tied down to a tree, the dog was being eaten alive. Getting up at once, I quickly released him, and he ran like mad down the muddy shore, splashing into the river. Less than a year old, he was playful as a puppy. While I was petting his dog, George emerged from his small tent, which was no more than a bug screen with a canvas roof but no floor. His face looked none too happy. "Damned mosquitoes!" he cursed. "I couldn't sleep all night."

Then, catching Cisko and retying him to the tree, he mumbled: "You can't get soft on a dog. All he needs to know is who's boss, that's all."

Yesterday, soon after we had started our journey, he had dealt Cisko a few blows on the head with the paddle to teach him to lie still and not jump about in the canoe.

Our breakfast of oatmeal and sandwiches was done and over with quickly. "Today, I'll get us some meat," George promised again.

After boarding the canoe, we paddled long and hard to get away from pursuing bugs until we could sit back and relax, just looking about. Much the same as yesterday, mile after mile of forested banks glided by. All we saw were one shy beaver and a few birds. When we passed a couple of ducks, George quickly raised his twenty–two and got off several shots without hitting the target.

"Can't aim from the boat," he said: "Next time I'll land first."

During the afternoon, we heard the loud calls of Canada geese ahead. This was the time of year that waterfowl moult their feathers and are flightless. Upon our approach, the wary geese clambered up the shore and disappeared into the forest, taking their chances with four–footed enemies. George never got another opportunity to shoot until we entered a stretch where the river widened. Far ahead, we spotted half a dozen geese loafing on a sand bar. George immediately beached the canoe, tied it down and got out. Carrying his twenty–two, he disappeared into the forest, intending to sneak up on his quarry. Presently, a couple of shots rang out in quick succession, but George returned without meat. "Next time I'll use my big gun," he scowled.

We soon spotted another group of geese, again on an open bar. Using the same tactics as before, George left me behind in the canoe and took off into the woods, carrying his high–powered rifle. Many minutes went by in silence, until two shots went off in quick succession, booming across the river and echoing against the hills. Presently, he returned without geese and without comment. After we had gone underway again, he muttered: "Damn mosquitoes! I couldn't aim."

~

Our second overnight camp was a repeat of the first. Same kind of food, same hordes of bugs. This time I tried out my head net, but found it hot and bothersome. To get away from the hated pests, we retired early. Lighting my pipe inside the tent, I hunted down the mosquitoes that had entered with me, until I could finally close my eyes.

During the next day, George mentioned that we had made very good time. One more overnighting and we should reach our destination. However, since we had plenty of time, he wanted to make a couple of stops along the way. "I'll show you some nice ponds a little ways off in the woods," he said: "Lots of moose there. We'll see some for sure."

Although he did not say so, his intentions were clear: he wanted venison, some for our immediate use and the rest to take home to his parents. Neither the prospect of watching a moose being shot and butchered, nor the idea of a canoe loaded with bloody meat, appealed much to me. Yet, I said nothing. And, as it turned out, we saw nothing.

The following morning was cold and rainy. We paddled steadily, not only to keep warm but also to cover more country. At one moment when we were both rowing strongly and creating a fair bow wave, George casually remarked: "This is how fast a bear can swim!"

So much for my assumption that we were absolutely safe on the water! Just suppose we surprised a grizzly close by on the shore as we floated by slowly, and the animal lunged after us aggressively…?

~

We interrupted our journey to walk a little distance into the woods to a marshy pond, where we spotted ducks, terns, and some songbirds, but no moose. At our second stop, by the ruins of an old trapper's cabin, we secured the canoe and made an early supper. On the opposite shore, a game trail led from the trees to a calm backwater. Here, according to George, moose often crossed the river. One day a grizzly had emerged instead and swam over. George had shot the big bear just as it had climbed the very bank on which we were now sitting. "He fell back with a big splash," he recalled with a grin.

Gesturing with his rifle, George pointed out at what distance he would be able to hit a moose, much farther than the "old–timers." When he was a kid, his uncle was still using his bow. Although its range was limited, it was very powerful. "Once he shot a big bull here. The arrow went right through his neck!"

After our break, we kept going well into the evening. The later in the day, George claimed, the better our chances of seeing wildlife. The river gradually widened, the current slowed down, and the shores became low and marshy. Eventually we entered the wide Nisutlin Bay. Nearby was some good moose habitat and this is where George intended to make our last camp. However, I had different ideas. The huge expanse of open water ahead looked like a very hostile place to be on a windy day. Right now it was calm, and the skies were clearing, ideal conditions to follow the shortest route, right across. George objected, claiming to be very tired: "I can't paddle all day and all night," he pleaded.

His real reason for wanting to stay another day was no doubt his desire to bring home a moose. When I insisted that we should take advantage of the favourable weather, he tried another excuse: "What day is tomorrow?"

"Sunday!"

"I don't want to be home on Sunday," he answered.

"Why not?"

"The priest will make me go to church!"

Recalling that the small hamlet of Teslin featured a Roman Catholic church as well two or three protestant houses of worship, I asked: "What do you think of the Europeans coming here to convert the Indians to several different religions, each claiming to be the only true path to heaven?"

"What do you think?" George countered diplomatically.

Tentatively, not wanting to hurt his pious convictions, I replied: "Oh, well, to me they are all the same."

"To me, they're all the shits!" dead–panned George.

That settled the matter, and with a laugh, we took up the paddles and headed directly across six miles (10 km) of open water under a luminous sky. Loons were yodelling their haunting calls far and near. Sometimes they sounded almost like a wolf. However, no matter how keenly I wanted to hear a real one, that chance slowly vanished the closer we got to the highway and town.

CHAPTER 2

WOLF SONG AND OTHER MUSIC

"What are you doing here on the river, all by yourself?" asked the old Indian woman. And looking askance at my heavily–laden canoe, she added: "Do you know there are rapids around the corner?"

I had just landed to pay a brief visit to her family camp along the Hootalinqua River in Yukon. The old woman, a large knife in her bloodied hand, sat on the grassy bank by a heap of salmon. While she chatted, surrounded by several children, she continued her work, expertly gutting a huge fish, splitting the carcass and placing it flat on the ground. Then, she made a series of deep incisions into the sides, half an inch (1.2 cm) or so apart. This was to facilitate curing of the flesh. Behind her, between the willow bushes, stood a six foot high (2 m) drying rack constructed of green branches on which the carcasses were hung to be smoked over a smouldering fire. The salmon had been netted the previous night by the men in the group, who had floated downriver in a pair of wooden, home–made skiffs, a net strung in between. The migrating fish, hurrying up the current to traditional spawning grounds, had been caught in abundance. Taking full advantage of the yearly but temporary bonanza, the family was working on its winter food supply.

"Just looking for wildlife, eh…?" The woman's broad, deeply lined face betrayed amused disdain as she repeated my answer to her earlier question. Crazy white man! To hard–working, pragmatic people like her, my simple desire just to see animals might have seemed silly. In those days, now some forty years ago, most other whites came here to hunt, fish, trap, prospect for minerals, or pan for gold.

"Aren't you afraid of grizzly bears?" she asked again. "Have you got a gun?"

And when I replied that I carried a twenty–two (which I had borrowed from George), her derisive grin broadened. A twenty–two against a grizzly!

"What are you going to do if you meet one? You better sleep light or a bear might chew your head off before you are awake. Hehehe…."

According to her son, who had just emerged from the tent, bears were not rare in these parts. You just never knew when you got to face one. "A prospector was seriously mauled here last summer. His guts ripped out with one swipe of its paw!"

One story followed the other as they usually do around campfires in the Yukon. Respect for the awesome and unpredictable grizzly was deep–seated among natives and whites alike. Most residents of the North with whom I had raised this ever popular subject would never go anywhere in the backcountry without carrying a high–powered rifle, let alone all by themselves for several days and nights….

The woman's admonitions did little to improve my already shaky confidence, not so much in respect to the bears, but about the water ahead. I had hiked and camped in grizzly country before, in the mountains of Alberta, accepting the risks involved and naively hoping for the best. But I was still a novice at canoeing and this was my first trip alone. The map showed that there were indeed rapids around the next bend. I could already hear their menacing roar. Drifting downstream on the narrowing river at increasing speed, I felt a choking sense of apprehension. Several kingfishers flushed from overhanging branches. Their shrill, staccato calls reverberated against the steep banks like a burst of invective, an omen of impending doom, or so it seemed to me. When the standing waves of the rapids came into view, like the manes of angry white horses, I began to paddle as if my life depended on it to get to the opposite side where the water seemed less rough.

To my relief, the little crisis was over in minutes. Steady as a rock, the canoe danced through the turbulence and glided out into the smooth waters beyond. Ahead lay miles of navigable river and the map indicated no other rapids for at least a day's travel. I began to relax, looking around with awe at the vast, pristine scenery unfolding beyond every bend.

~

My adventure had been planned the day before when I had met Leslie, a British immigrant, who had worked in a variety of trades and was now operating a fishing–guide business. He told me that a party of Indians had heard a pack of wolves a week or so ago at a point where Hundredmile Creek entered the river, about thirty miles (50 km) downstream. The natives had shot a moose there, and while they were butchering the animal, wolves had howled from both sides of the river.

To me, the wolf represented the embodiment of the spirit of the wild that I had come to find. Thusfar, the legendary predator had eluded me, here in Yukon as well as in Alberta, and my desire to see and hear one had only become stronger. My hopes high, I had started out this morning from the river bridge on the Alaska Highway. The canoe had been lent to me by a generous member of the Whitehorse canoe club, and Leslie had given me a beautiful Husky bitch along for company. I had accepted his offer, not only because the dog was very friendly, but its weight would help stabilize the slender, seventeen–foot (5.1 m) craft.

I soon realized though that my four–footed companion required constant supervision. If she turned her heavy rump to the left, I was forced to shift my weight to the right as a counterbalance. It happened so often that I began to suspect that she was playing games with me. Huskies are supposed to be very smart. I had read about the sneaky tricks these dogs, when harnessed to a toboggan, can pull on their human masters. This had to stop! After all, who was supposed to be in control here? What if she decided to stand up and lean overboard, perhaps if she spotted a rabbit on the bank? So, following the example of my native guide on a previous trip, I taught the dog to lie still with a few light slaps of the paddle. She obliged with a sly wink in her wolfish eyes. All I had to do next time we neared the shore for a landing, was to keep her from jumping out too eagerly and upsetting the canoe… or from jumping back in again all over the gear, her big paws heavy with mud.

Leslie had assured me that the river would provide an easy ride down. In wide, lazy turns, following the path of least resistance through the mountainous country, the Hootalinqua was a natural travel route regularly used by native and non–native inhabitants of this region. In recent years, the amount of traffic has greatly increased, but in 1961 one could drift down for days without seeing another human. The shoreline woods were still completely wild and untouched until one reached the next bridge on the Dawson highway, several hundred miles away. I was not going that far, and as to my way back upriver, I had hired Leslie's services to pick me up in his powerboat five days from now.

It was a warm summer afternoon with very little wind, and toward evening the languid atmosphere became still as in a dream. Crowned by an arch of distant mountains under a clear sky, the river mirrored the tall spruces on the bank. Drifting along silently on water that was as smooth as glass, I had the impression of being suspended between two identical wonderlands, one of them upside down. Below, through the crystal–clear shallows, I could see the bottom slipping steadily by. The effect was dizzying.

The current was often less than two miles (3 km) per hour, and most of the time I kept up an easy paddle to boost my speed, using the traditional Ojibwa or J–stroke, which I had quite recently learned from another canoeist. It had made a watershed of difference in handling the responsive craft. Instead of zigzagging along, I could now follow a straight course. By twisting the blade more or less inward and forward at the end of each stroke, I could correct the drift or even turn into the opposite direction while paddling on one side only. If my rowing arm became tired, I switched over to the other side. This, to my mind, is the great advantage of the Indian canoe over European canoes and kayaks which are powered by double–bladed paddles and require the simultaneous use of both arms.

My destination for the trip was marked on a large–scale map that showed every bend and island, allowing me to check on my progress. Just over the midway point was Canyon Creek, which featured a scenic waterfall where Leslie had often seen otters. There was also good fishing nearby, another tempting reason to make an overnight stop. I saw no otters but managed to catch a couple of graylings for supper. When I threw the heads of the fish to the dog, he swallowed them in a trice.

Next day, after another six hours of easy travel I finally arrived at the mouth of Hundredmile Creek. There was a wide gravel bar and plenty of level space in the woods, ideal for camping. However, I decided against it. This was supposed to be the location where the Indians had butchered their moose a week ago. The discarded offal—guts, hide, head and legs—could still be around, which might attract unwelcome visitors during the night....

Checking out a muddy stretch of the gravel bar, I indeed found fresh bear tracks. The imprint of claws was well ahead of the toe pads which identified the animal as a grizzly, a large one. Glancing around uneasily at the thick bush on shore, I continued my investigation and located other tracks that were much more appealing: wolf! There were huge adult–sized prints of nearly five inches (12.5 cm) wide, as well as the much smaller ones of pups! It looked like a whole pack had been here some time ago. Hoping that they might return, I planned to spend a few hours here each evening over the next three days, but to make camp, I went elsewhere.

~

Along these boreal rivers, perfect campsites can be far and wide between. Half a mile or so back, I had noted an open cutbank which I reached by paddling against the current and, like a beaver, taking advantage of the eddies close to shore. After dragging the canoe onto the mud below the bank, I carried my gear up while the dog happily explored the woods. Although the high ground was bumpy and uneven, there was a level spot just large enough for the small tent, which I placed on a bed of spruce bows. I then organized a handy log to sit on, with a wide view of the river, and made a fire for a quick meal. Afterwards, there was still plenty of time left to go back to the gravel bar by the creek. Leaving the dog in camp, tied to a tree, I hurried on my way. The food box and a heavy rock served as counterweights in the bow of the canoe.

Across from the mouth of Hundredmile Creek, the river speeded up over shallows where salmon were spawning. Every few seconds, a large bright–red fish broke the surface or jumped clear of the water. Commonly called King Salmon, they had come from the Bering Sea about 2,000 river miles (3000 km) away. In one of the miracles of nature, these fish had fought their way up the watershed to reach this site, the very locality where they hatched four or five years earlier. Here, they were to perpetuate their species and destined to die. Having left the ocean as silvery fish, their scales had been recycled by their own bodies for energy, and their colour had flushed to blood–red. The beaks of the males were deformed into a hooked, nuptial grimace. In a frenzied mating ritual, the salmon dug craters in the gravely river bottom that were one to three feet (30–90 cm) deep. When finished, the females exuded some 4,000 eggs each and the males, shuddering in ecstasy, released clouds of milky sperm. Their task completed, the spent fishes would quickly deteriorate and die, washing up onto the shores, fertilizing the waters or becoming food for a host of scavengers.

A bald eagle watched from a snag as I parked the canoe on the opposite shore in the shade of some overhanging spruces. Ravens and gulls flew by searching for carrion. Preparing for a long vigil, I kept binoculars and camera at the ready. But my wait proved to be in vain.

~

Returning to camp just after sundown, I was welcomed by the dog with tail–wagging enthusiasm. Although Leslie had assured me that she would not need any food at all for five days, I threw her the head of a hare I had shot and gutted along the riverbank nearby. It was gone in one gulp and she still looked as if she were starving. She was however quite fat and in good condition. Perhaps she was even catching her own hares. Each morning and evening after I released her, she disappeared into the woods like a shot, panting contentedly upon her return.

Hares, or snowshoe rabbits as they are usually called, happened to be at the peak of their population cycle. During evening, it was common to see several of the cute bunnies along the forest edges, so I stopped feeling bad about having shot the first one I saw for the pot. Soon enough, the population was bound to crash, most of them succumbing to starvation or disease, or fall prey to the increasing number of predators.

Hawks were common. One morning I heard the sharp click of claws on wood in the crown of a tree above. There, staring down with his fierce, red eyes, sat a goshawk. This powerful raptor of the forest locates its prey by sound as well as sight. So does the equally ferocious great–horned owl. At dusk, while I sat smoking a pipe by the fire, one of these nocturnal hunters looked me over from a sapling by the tent. It felt good not to be a defenceless small creature such as a rabbit! But then, there were moments when I felt hunted too, especially at night fall....

The thought of prowling bears never really left me, but by the second day in the wilderness I had reached a more relaxed state of mind and shrugged off undue worry. The natural beauty and silence that surrounded me, the complete lack of man–made noise and clutter, acted as a balm on my overworked imagination. Yet, hoping to deter four–footed trespassers, I left olfactory evidence of my human presence all around camp, like a dog marking its territory. Upon retiring into the tent, I took the loaded twenty–two along. The first night, thinking that noise might scare off possible intruders, I even kept an aluminium pot handy with metal utensils in it. Fortunately, the need to test these rather paranoid defences never arose.

If I awoke during the night, I listened to every rustle, bump, and crackle. From across the river came the sucking noises of moose wading in the shallows, water gushing from their heads as they pulled up strings of aquatic vegetation. The giant herbivores were hidden by darkness, yet I knew what they were doing as surely as a lynx or fox knows about the mice rustling under dry grass. Hearing becomes a new sensation in the wilderness. It is sharpened and ever alert.

After the moose had gone back into the woods, the stillness of the night was occasionally interrupted by the splash of a passing salmon or the hooting of the great–horned owl. Even after all animal sounds stopped, the silence never seemed complete. My ears were filled with an high–pitched din that is said to have a physical cause: the pressure of blood rushing by one's inner ear. There is even a medical label for this: tinnitus. But to me it was like distant music, its melody urgent but confused, like the tuning up of a thousand violins, an orchestra of astronomical size preparing for a majestic symphony. Perhaps, as some people claim, such eerie noises are caused by the northern lights, by electromagnetic vibes high in the sky. Spectacular displays of *aurora borealis* were a frequent phenomenon at this latitude but to see them I had to venture out of my cosy, triangular cave. And that, admittedly,

I was loath to do. No matter how thin the canvas, it gave some sense of security. No wild animal would ever disturb a sleeping human, or so I wanted to believe. And if it came to pass, there was nothing a mere mortal could do about it....

~

In the chill of early morning, all thoughts of danger and mystery were extinguished with a few handfuls of cold water in the face. Feeling reborn, the first order of business was to start the fire and hang a pot on the tea stick. After breakfast, I passed the long day exploring oxbows and backwaters, or looking for animal tracks on muddy points. However, during the heat of afternoon, the bugs spoiled much of the pleasure, harassing man and beast alike. It was now late summer and the mosquitoes had declined in number, but other pests were at their worst. The big horse flies could deliver a jab that felt like a red–hot nail going in. Much smaller, less than one–third the size of an ordinary house fly, black flies bit tiny holes in one's skin that could fester and cause welts itching for weeks. The smallest species of black fly, locally known as "no–see–ums" were the most numerous and could be hardest to cope with, since you did not feel them on your skin. They often went to work behind your ears or in the neck.

However, if I thought I had reason to complain, what about the animals? A moose, standing in the shallows, was buzzed by angry swarms of thousands or perhaps millions of flies. When it swam across the river, the poor animal was followed by its vicious, blood–sucking tormentors, literally hanging like a cloud over its head.

By early evening, I sought relief from the bugs on the river and canoed down to Hundredmile Creek. There, I waited in the shade of the overhanging spruces opposite the gravel bar. On the very last evening, in the calm before sundown, my fondest wish finally came true. I heard a wolf!

The animal must have been miles away for its call was no more than a barely audible, melancholy wail that seemed to express all the world's sorrow. It died away again in seconds, merging with the drone of insects and the hiss of swift water over the shallows. In fact, the experience had been so brief and surrealistic that I began to think that I might have imagined the whole thing. But then, the wolf howled again, three times, a few minutes apart.

I kept listening for a long time. A dragonfly hunted mosquitoes around the canoe, the rustle of its brittle wings surprisingly loud. Occasionally, a salmon shattered the glassy surface of the river, the low sun flashing on its scarlet body before it crashed back in sparkling spray. But the wolf did not call again.

That evening I huddled by the smouldering fire in a contented trance. As the silence of night closed in, the haunting voice of the wolf resounded in my ears. Later, lying in the tent on the verge of sleep, its howls seemed to be real. Or were they? I opened the bug screen and stuck my head outside. The sky was bright with splendour. Banners of northern light waved overhead, veiling the stars. Fluorescent columns and curtains pulsed on the horizon. A crescent moon, low over the humped blackness of the forest, smiled at its inverted reflection in the river. I lay looking and listening until shivers ran down my spine and made me retreat again into the warmth of the sleeping bag.

When I woke up for the second time, it was dawn. The stars and aurora had gone. A pale mist hung over the water. The wolf was howling again, and this time it was real and much closer than before. Its deep, sonorous baritone echoed across the river. As the sun rose, shrouded by fog, a second wolf joined in, and another, and another, their high voices richly modulated or shrill like puppies. Their chorus climaxed into a frenzied cacophony that stopped as abruptly as it had begun.

I got out of the tent. The sun had burst through the mist, its rays glittering on frosted spruces that reached up to the heavens like giant candle flames.

The dog appeared equally impressed and subdued as I. Upon release, she did not stray from my side. She lay nearby when I built a fire and she followed on my heels whenever I got up to collect wood. Only she knew how close her wild ancestors had been around our camp that night.

CHAPTER 3

FISH, FATE AND FAREWELL TO YUKON

Eighty–eight years old and now practically blind, Sten had spent most of his life in the bush, trapping furbearing animals. "If only I had had a camera," he mused: "I would have had some pictures to show you."

When I first met the old Swede, he was sitting in the sun on the boat launch in Teslin. Our chat quickly turned to fishing. If I would give him a hand getting his boat into the water, he would take me along for a spin in the bay.

After he had started the small outboard and slowed down to trolling speed, he handed me a crude fishing line, wound on a two–foot (60 cm) plank. The lure, fitted with a single large hook, was made from a strip of bare tin about eight inches long by two wide (20 x 5 cm). The one he used was almost twice as big. I never thought he would catch anything at all with that! However, in half an hour he pulled in two lake trout, the biggest weighing about ten pounds (4.5 kg). I caught nothing.

"Well, that'll last me a few days," he said. "Let's go home and make supper."

While I pulled the boat back on shore, he cleaned the fish, throwing the offal to some of the thievish dogs that roamed the town. (That morning, a tiny cur had jumped into my car through the open window and taken off with an entire Edam cheese!)

It was only a few steps to his house. The kitchen was spartan, messy, and overrun with mice. Without paying any attention to his saucy boarders, the old man fried a portion of the fish in a large skillet and boiled a pot of unpeeled potatoes. Perhaps he relied mostly on routine, or perhaps he could see reasonably well at close range. At any rate he did not ask for my help, and when all was done, he placed a pot of butter in front of us on the table.

It was not the most appetizing meal I had ever enjoyed, but his stories were entertaining. Like most people who spend much of their time alone, he babbled on incessantly, giggling about the vagaries of fate in the bush. At fifty he had had enough of his bachelor existence and married a native girl of sixteen. She had left him years ago. "I figured she thought I could not give her any more what she wanted," he chuckled.

When we parted he repeated a suggestion he had made before: "If you want to see grizzlies and other wildlife, you take the Big Salmon River. A little ways down, at Scurvy Creek, there is a big spawning ground. That's where you'll find the bears for sure later this summer."

And, as if he suddenly thought of something that had happened to him once, he warned: "But watch out for sweepers and log jams…!"

The idea of canoeing down the Big Salmon River stayed in the back of my mind. But his parting admonition was soon forgotten. In fact, its significance never really registered on my greenhorn mind at all until it was too late.

~

For much of that summer, apart from the enjoyable floats down the Nisutlin and the Hootalinqua, I explored the Yukon by road. I drove to Haines on the Alaskan coast and to the remote Klondike goldrush town of Dawson City. Along the way, I made a few day trips on lakes and creeks. My tiny canoe proved to be very steady once I got the hang of it. However, for another long river journey I needed a larger craft and a partner. Campgrounds were always good places to meet people. One day I struck up a conversation with a guy who was carrying a canoe on his small truck. He introduced himself as John, a dentist from Montana. Tired of the rat race, he had escaped for an extended holiday to Alaska.

John would not mind making a river trip with me. When I mentioned the Big Salmon and the Stewart, he studied the map and saw that both rivers were tributaries of the mighty Yukon, which flowed by Dawson City, a place he wanted to visit anyway. While the Big Salmon looked too narrow and difficult for his liking, he noted that the distance to Dawson was shortest via the Stewart, about 250 river miles (400 km). Thus, our choice was easily made.

After driving both vehicles to Dawson, we parked my Pontiac and drove back in John's truck to the starting point by the highway bridge across the Stewart, which allowed easy access to the river. Swollen with recent rains, the current ran high and dirty at a speed of about eight miles (12 km). Now, in retrospect, I shudder just thinking about the chances we were taking. Although it seated two persons, John's canoe was just a shallow pleasure craft. By the time we had loaded up, there was no more than a few inches of free board in the centre, and my companion, like me, had very little experience with river travel.

Fortunately, we did not encounter much in the way of rapids. Even the wide and windy Yukon allowed us to proceed apace without mishap. Along the way we traversed some beautiful, park–like country with open hillsides on which we saw several black bears, far and near. We even spotted a wolf, the first one ever for both of us! Striding regally along the shore, the long–legged predator casually glanced at us as if we were a piece of driftwood floating by. Seconds later, he vanished into the trees.

~

It was not until late August that I met Wayne, a young fellow from Winnipeg, who had come north to hunt and fish. Although he too was a novice on swift water, he had brought a superior canoe, a sixteen–foot (4.8 m), cedar and canvas Chestnut Prospector, extra wide and extra deep. Cancelling his own plans for the remainder of the summer, he was keen to descend the Big Salmon with me as soon as the weather cleared.

It rained for a week and by the fifth of September two inches (5 cm) of snow had fallen. Nevertheless, tired of waiting, we chose to push on despite the lousy weather and muddy roads. First we drove both cars to our planned destination where the Dawson road crossed the Yukon River. We then took one vehicle back to the starting point on Quiet Lake along the trail to Norman Wells. By the time we finally arrived there, the sky had cleared and the night was bright with stars. Excited about tomorrow's prospects, we crawled into our tents, ready for an early start next day.

However, upon awakening, we were in for a nasty surprise. A blustery wind had sprung up, driving huge waves onto shore, crashing like ocean surf. Low clouds scudded down the valley, shrouding the mountains and soaking us with intermittent showers. After breakfast, we decided to go for it anyway. The high wind would be in our backs and push us into the direction we had to go, eight miles (12 km) down the long lake. Folding our wet tents and packing our gear, including enough food for about two weeks, we covered the precious cargo snugly with a tarp. Then, foolishly disregarding the ominous conditions, we started off… and collided with a metal rod jutting out from the boat ramp just below the water line.

"There is no damage, eh?" shouted Wayne from the stern.

"No way," I answered: "We barely touched the thing."

Just to make sure, I ran my hand along the canvas behind me near where the rod had hit, and I found a hole that fitted my finger, right through the canvas and wood.

Returning to shore and unloading the canoe, Wayne plugged the leak with some cloth glued in place with resin that we collected from spruce trunks and melted in a can over a fire. In the meantime, I was wringing out my socks for I had suffered the misfortune of falling through the rickety boat launch and filling my rubber boots with water.

The weather was deteriorating and it looked like the rain was settling in for the day. Ignoring the string of evil omens and hiding our misgivings, we boarded the canoe for the second time and headed out into the waves. Instead of staying close to shore and working our way from one peninsula to the next, we made for the shortest route, going with the wind. Far out on the huge lake, the waves were about five feet (1.5 m) high, alternately lifting the heavy canoe and sucking it down into the trough of the next building breaker. Surging ahead, rising and descending in a mesmerizing rhythm, we were flying down the wind like Norsemen on the Atlantic Ocean.

The far shore was a line of spruce. Having studied the map, we knew that the outlet creek was somewhere in a bay in the right hand corner. By the time we got close, we ended up a little too far to the left and had to tack into the wind. Choosing the best possible moment to turn across the breakers, between two gigantic crests, I shouted: "Now!"

As the bow of the canoe was lifted high by the second wave, my paddle hit nothing but air. Losing my balance, I almost fell overboard and the canoe came close to capsizing. Saved from disaster by the skin of our teeth, we rode the furious onslaught and slowly fought our way upwind, each wave sloshing water inside, until we once more had to switch direction. "Now!" Swinging the canoe downwind, we rushed into the bay, pushed by the swell. What a relief! The rain had stopped and between the shredded clouds a few radiant beams of sunlight were bursting through, like a blessing from the Gods.

As we proceeded into the creek, the canoe was gripped by a strong current surging over a gravel bar, which we barely cleared. The channel narrowed, our speed increased, and we needed all our wits to dodge boulders that blocked the way. Twice, one of us had to jump out, holding the canoe in check. Soon, the water became so shallow that we both had to get out and proceed by wading alongside. During a brief stop to catch our breath and eat a snack in the lee of an island, Wayne put on his hip waders and suggested that we wet a line. The fishing proved to be absolutely fantastic. At every cast, our flies were struck by graylings. Soon we had pulled in half a dozen, enough for a meal.

Pressing on, we negotiated the remainder of the turbulent channel, partly floating and partly wading beside the canoe, until we reached Sandy Lake. It proved to be a stunningly beautiful body of water rimmed with tall, dark spruces, offset with the golden foliage of poplar and birch which were already beginning to turn colour. The clouds slowly lifted and revealed mountains which had received a fresh coat of snow.

A sandy beach afforded an ideal campsite. In high spirits we unloaded the canoe, set up the tents and built a fire. While one of us fried the fish to a golden brown treat, the other hung our gear to dry in the warm sun. In the evening, we strolled along the beach like conquerors, feeling very content and pleased with our perseverance. The clear sky seemed to promise that Indian Summer had finally arrived, a favourite season in the North, characterized by warm days, cool nights, and a respite from the summer's insect plague. We were counting on at least two weeks of fine autumn weather, ideal for making the ambitious wilderness trip on which we had embarked.

However, as if the devil were playing games with us, our luck changed again overnight. Waking to the drumming of heavy rain on canvas, we did not even bother getting up until noon. The wind had turned north and our beach was washed by waves. Keeping an eye on the rising water level, and trying to cope with a leaking tent, we spent a long, dreary day and night in camp.

Next morning, the rains had stopped. All around, the wooded foothills were cloaked in snow, starting just a little above the elevation of Sandy Lake. Although the wind was still strong, by the time we had finished breakfast, the clouds had lifted somewhat and a few patches of blue appeared. Once more our hopes surged. Filled with happy anticipation, we secured our gear in the canoe and covered everything with a tarp.

Following close to the wooded banks, we had no trouble reaching the lake's outlet on the opposite shore. Here, the country was quite marshy and we saw many fish–eating birds such as mergansers and kingfishers, as well as the odd osprey or bald eagle. At the sight of ducks flushing ahead of our approach, Wayne took his shotgun out of its case. As we rounded a low point studded with willows, several mallards took wing against the wind, then veered back quite close. Wayne downed one of them with a single shot. He was apparently a keen hunter of waterfowl as well as a fisherman. Collecting the duck from the water, he grinned: "This will add a bit of variety to our diet."

A few miles farther down, the creek emptied into Big Salmon Lake, the third and last sheet of open water we needed to negotiate before reaching the river of the same name, a few miles farther west. The lake was less than one mile (1.6 km) across but the view of what lay ahead of us was far from inviting. The wind, still very strong, had turned to the northwest, blowing down the length of the long valley. The surface of the lake was a raging sea of whitecaps.

"How are we going to cross that?" mumbled Wayne.

"Let's think this over for awhile," I responded, steering the canoe into the lee of the bushes before leaving the creek.

After a pause, I added: "Once we get to the other side, we should be okay. Considering the wind direction, the opposite shore may be a little more sheltered."

Wayne said very little. He had switched to the front seat, a more favourable position for fishing and shooting, and he was pensively plucking his duck, feathers blowing on the wind. It was too cold to stay here long and the nearby shores looked far from inviting as a place to camp. When Wayne was finished, we set off in silence.

Angling a little away from the wind, even though this was not the shortest route, we soon had only one thought on our minds: just to make it across and get back to the safety of land. The farther we got, the more powerful the waves, each crest slopping water inside the heaving canoe. Soaked by spray, Wayne was straining at the paddle as hard as I. Watching him rise and fall made me feel sick to the stomach. Looking away, I noticed that the western sky had turned black. The wind had increased to a gale.

"We are not going to make it!" Wayne cried out.

Sharing his doubts, my throat dry with terror, I did not respond. Ahead and behind us, the shores seemed equally far. What if we were to be swamped here in the middle of the lake? No one would ever know! The concept struck me with explosive clarity. This was the reality of the wilderness I had sought! Cold, unforgiving, hostile, fiendishly inhuman. If we got into serious trouble, there was no hope in hell that anyone would rush to our rescue.

In desperation, I turned the canoe downwind. It reduced the violent pounding we were receiving. We could now increase our speed and almost keep abreast of the waves, the heavy canoe surfing on the crests. We were carried far off course, and gradually I steered a little closer to the wind, nearing the far shore. However, my hopes of finding quieter waters were dashed. The entire lake was in uproar, making it impossible to attempt a landing without crashing onto rocks. Drifting farther east, we finally rounded a point that sheltered a shallow bay where the waters were a little less turbulent.

"I never thought we were going to make it," said Wayne a while later, as we rested our tired arms.

"What now?"

In retrospect, I am sure that anyone with an ounce of sense would have chosen to go on shore to wait out the weather, no matter how dense and unappealing the woods might have looked. But the rain had stopped, the cloud cover was breaking up, and once more our courage began to bounce back. Hope springs eternal, at least to dreamers like us.

Hugging the shore, we went back around the point, testing the waters. Slow and steady would do it, I thought. We had seen worse. For one thing, heading west we could nose right into the waves and the canoe had proved to be very sea worthy, riding out the biggest of rollers. For another, the nearness of the shore felt reassuring, so very much better than the sheer panic we had experienced in the middle of the lake. We had roughly four miles (6 km) to go. Taking advantage of the lee behind every point and island, we began to make fairly good time. Moreover, the wind was abating, and the farther west we got, the smaller the waves. Eventually, it became quite calm. The sky overhead had cleared and the remnants of clouds were dissipating over the mountains. What a view! What a splendid country!

"The risk we took!" I mused: "If only we had waited a few hours..."

Wayne chuckled. He had already forgotten our peril of the past few hours. His thoughts had returned to fishing. Placing his spinning rod in the bow and letting out line, he resumed paddling at a slow trolling speed. Minutes later, while we crossed a bay with an inviting sandy beach, something struck his spinner.

"Got one," shouted Wayne: "A big one!"

His rod nearly bent double, he worked for some twenty minutes and still the fish had not broken the surface. Each time, after he had slowly reeled in much line, he lost it again in a second as the fish dived back. It was in fact pulling the canoe. When it was at last played out and we saw its huge, speckled form shimmering in the clear water, Wayne cried out: "A lake trout! That thing is at least twenty pounds (9 kg)!"

Far too large for our landing net and too heavy to lift by the line, how would we ever get this monster into the canoe? Only one prong of the triple hook appeared to be holding and it was badly bent.

"Shooting! We got to kill it first!"

Wayne's idea proved foolish, even illegal as he admitted later, but I grabbed the twenty–two. The small bullet went right into the gaping mouth of the fish. In the next instant, it flipped its tail and was gone, the reel screaming. Then, the line suddenly went slack and Wayne mumbled sadly: "Lost it..."

"No, wait!" Reeling in fast, he found the weight on the end of the line far less than before. "He must be dead."

To our astonishment, when the fish surfaced, it proved to be smaller than the first, no more than a pound or two (0.9 kg), but very much alive. It was easily netted and lifted aboard. What had happened? Apparently, in the few moments that Wayne had lost the big fish and began retrieving his spinner, another lake trout had struck. It certainly was not our proudest moment, but it illustrated the fantastic fishing opportunities on Big Salmon Lake in those days.

~

We beached the canoe in the sandy bay and prepared a gourmet dinner of trout and duck, barbecued on green sticks over hot coals. Time was flying by and we needed to push on quickly if we wanted to reach our goal for today, the mouth of Scurvy Creek, recommended by Sten the trapper. Studying the map we saw that it was no more than six miles (10 km) down the Big Salmon River, which drained out of the lake around the next point.

Excited about the improved prospects, we quickly stowed our gear back into the canoe, not bothering with our usual safety routine since wind and water were dead calm and we did not have far to go. I did not secure my most prized possession, the 10x50 Zeiss binoculars, with a piece of string to the boat, as had been my standard habit to prevent its loss if accidentally dropped into the water. Instead, I just put the glasses in the backpack. In my hurry to close the pack, I broke the string and left the top open. In addition to the binoculars, the pack also contained my two cameras, a long telephoto lens, a lantern, and other odds and ends. I did not tie the pack to the gunnels, nor did we secure and cover the cargo with the tarp, as had been our habit during the past few days. Little did we know that these seemingly trifling matters would become so crucial and so soon....

The waters of the Big Salmon River began their long journey to the mighty Yukon and onwards to the Bering Sea in a very pretty bay, its shore bordered with rushes and studded with willows. Beyond stood a wall of tall, brooding spruces. Like a regal sentry, a bald eagle sat on a snag by the water. A flock of colourful mergansers rushed away as we entered the narrow but deep outlet. A school of graylings stood in the clear current, and Wayne could not resist "to wet a line" as he was fond of saying. While he was happily engaged, I watched a kingfisher and a sharp–shinned hawk play a game of tag. The spunky raptor, not much bigger than its playmate, pursued the kingfisher vigorously back and forth. It looked as if the hawk was deadly serious, but as soon as it quit the chase and perched on a branch, the kingfisher returned to ask for more. Scolding loudly, it perched close by the hawk, until the latter took up the game again.

Suddenly, a goshawk, the sharp–shin's much bigger cousin, shot out of the woods and chased the kingfisher in earnest. Dodging the attack, its intended prey dived repeatedly headlong into the water. As soon as the hawk gave up, the kingfisher fled out of sight, indignantly sounding its harsh rattle.

The graylings proved hard to hook. Wayne stowed his rod in the bow, and we resumed our way downstream, quickly leaving the marsh behind. In the woods, we glimpsed a log cabin, its roof visible between spruces. Very soon, we had no more time to look around. The fast current required our undivided attention. As Sten had mentioned, this was indeed an important salmon spawning ground. The gravely bottom was pockmarked with holes in which the fish had laid their eggs some time ago. A few carcasses floated in the eddies near the bank. No wonder this was a favourite hang–out for bears!

Behind a sharp turn, I saw that our way was blocked by a sweeper, a large spruce that had fallen into the water, its roots undermined by the river. The flexible tip of the tree was alternately bending downstream with the current and swishing back. On the opposite side, a gravel bar jutted out into the stream, one side piled high in driftwood logs. In moments like this, when a canoeist approaches a difficult passage, good judgement as to how to proceed and the ability to make instant decisions, are of critical importance. My response and steermanship proved to be flawed.

As the gap between the top of the tree and the gravel bar did not seem quite wide enough to allow us to pass without scraping bottom, I thought it best to land somewhere, line up the canoe and walk it through the trouble spot. This was certainly the safest procedure. Deciding at once, I steered away from the gravel bar toward the willow–fringed shore, a little ways above the sweeper. Wayne was to grab hold of an overhanging branch and hold on, forcing the canoe to swing around and point into the upstream direction. This routine manoeuvre would have allowed us to land or to paddle back some distance and cross over to the opposite shore. Fate decided otherwise.

Just as the canoe was broadside in the current, Wayne lost his grip. That same paralyzing instant, I recognized our predicament, certain that this was the end. The canoe was carried downstream, slammed against the tree and came to rest against the branches. Then, in slow motion, it rolled over, spilling all of its contents, including us.

I tried desperately to pull myself up onto the horizontal tree, but it proved futile as my legs were in the grip of the powerful current. Letting go seemed risky but the only way out. Swept under and down in an instant, I surfaced in the deep pool beyond. Swimming, I ended up in an eddy by a steep, muddy bank. Clambering up, my eyes registered the fresh tracks of a grizzly, but my feverish mind was focused on the predicament we were in. All was lost, including the valuables in my pack sack: wallet, passport, car keys, and waterproof matches! And Wayne? Where was Wayne? Over the rushing of water, I heard his cries for help. He was hanging onto the tree for dear life, just his head above the raging torrent. Without hesitation I worked my way up on the prone trunk and helped pull him out. When we stood side by side on the shore, he shivered: "My waders… I still had my hip waders on. They were full of water. I didn't dare to let go!"

The canoe was still in the same place, lodged against the branches. I cannot remember how we managed to free it. All I recall, when we turned it upright, was the jolt of relief to see my pack still inside. All heavy contents, such as cameras and binoculars, had fallen out since the sack had been open, but the passport and car keys were tucked safely in one of the pockets! The very fact that I, in my hurry to get underway, had broken the closing string had prevented the pack from sinking and becoming lost.

One of the paddles was also inside the canoe. The other one had ended up the same way as I, into the pool. We found it lodged against the muddy bank with the grizzly tracks. We also recovered our sleeping bags. But not the tents, nor Wayne's shotgun, his brand–new Husqvarna rifle with the telescope sights, his movie camera, and all of his fishing tackle. In addition to the heavy objects that had fallen out of my pack, I had lost the borrowed twenty–two and my fishing rod. Also gone were our cooking pots, axes, lanterns, extra clothing, as well as all of our food, except some soggy packages that had stayed afloat inside the canoe.

At the very least we were still alive and in a position to get out of here easily. Recalling the cabin in the woods near the inlet, we at once decided to go there. Dragging the canoe against the strong current, we walked most of the way through the water, often sinking to our middle as we stumbled over the uneven bottom worked over by the salmon.

By the time we spotted the gable roof between the trees, it was dusk. To our relief the cabin was not a complete ruin such as several others we had seen along Yukon's waterways. Even the windows were intact. The door was unlocked. Looking about in the dark interior, it was plain that the place had not been used for some time. And everything was soaking wet, because the roof had leaked.

Our first concern was to start a fire in the rusted barrel stove. It turned out to be maddeningly difficult. There was no firewood at all in the place, nor could we find much outside. The nearby spruces had been stripped of their lower branches, and the willows were still soaking wet after the recent rains. Dry wood was at a premium. Trying to light the few sticks we had managed to collect, we wasted a good portion of our supply of matches. Giving up, we forced ourselves to go back to basics and start all over. Tearing a log from the ruins of a shed, and using a rusty saw and a blunt axe that we found inside, we cut and split some blocks. The insides were dry. With our knives, we made curly shavings and built a little pyre in the dark, wet stove. But the flame died soon. Repeated and increasingly frantic efforts produced no success either, until Wayne located a rusty lantern which contained a remnant of combustible fluid.

At long last, we could warm ourselves, hang wet clothes to dry, and heat some water. Our total food supply was one mushy loaf of bread, one package of pancake flour, one envelope of instant soup, and two tea bags. The flour proved useless, probably because it had become wet.

When we were ready for sleep and inspected the beds in a dark corner, we found the mattresses soaking wet. Throwing them outside, we laid down on the bare metal springs in our partially wet bags. Sleeping fitfully, I got up several times to attend to the stove, which we wanted to keep going all night. By now it was nice and warm inside, and willy–nilly I began to enjoy the adventure. It sure felt great to be in a cabin in the wilderness. How wonderful it would be to be here in winter, I thought, imagining wolves howling from the frozen lake. In the daytime, one could roam at will and look for their tracks.

Returning to bed, I wondered who had built this place. Sten the Swede? There were several bundles of traps hanging on the wall, and two rusted rifles stood in a corner. It looked like the place had been abandoned in haste. Why? Had the trapper perhaps come to grief in the wilderness…?

~

In the light of morning I discovered why I had not been able to get the putrid smell of garbage out of my nose. The folded canvas I had used for a pillow was covered with filthy mould. We also found out why our soup had tasted so foul last night. The inside of the kettle in which we had boiled water was coated with a thick layer of grime. As a substitute, we found a rusty tin. Punching two small holes in its rim, we inserted a piece of wire for a handle.

After the cool night, the day dawned bright with hoar frost on every branch. It looked like Indian Summer had indeed arrived. The sun was gaining strength quickly. We decided to stay over another day so as to dry out our sleeping bags and other things. Tomorrow we would make our way back to the road via the chain of lakes, the same way we had come.

During the afternoon, we heard the sound of an engine approaching from far away. Presently, a Zodiac raft came into view, carrying two men, one of them a native I had met before. It was Frank, the friendly and reliable guy who had driven my car back to Teslin at the start of my trip on the Nisutlin. He told us that he had been hired as a guide for the other man, a hunter who wanted to shoot a grizzly here. Along their way up, early this morning, Frank had killed and butchered a moose. He had left the carcass in the bush and intended to retrieve the meat on the way home in a couple of days. When we asked Frank where he had shot the moose, he readily told us: "Just this side of Sandy Lake."

In turn, the two men inquired as to the exact location of the accident in which we had lost such a long list of valuables. They restarted their engine and waved goodbye. As we watched them disappear, we could not help feeling a little envious.

To our surprise, our visitors returned in the evening on foot, guns cradled in their arms. They had walked along the shore of the creek, starting near the point where our mishap had taken place. Their intention was to salvage some of our more valuable gear from the deep water behind the sweeper. They had come to the cabin to scrounge material for a makeshift grappling hook. They also took the axe. All we asked for and got from them was a handful of matches.

The weather of the following day was a repeat of the previous one. Feeling very good, with dry clothes and sleeping bags, we embarked on our return journey under calm conditions. Along the way, we enjoyed one of the most memorable meals of a lifetime, courtesy of Frank. Mindful of what he had casually told us, we were immediately alert when ravens flushed out of the bushes near Sandy Lake. Making a lot of noise so as not to surprise a scavenging grizzly, we investigated the place and found the carcass of the moose. It had been partially skinned and the hide pulled over the meat. Everything was covered with the excrement of ravens. Wayne selected and cut out a portion of the tenderloin, while I wrote a note of thanks to Frank, attaching it to a branch.

We made a fire on the same pretty beach where we had spent a weather–bound day and night on our eventful way up. Boiling water in the billy can, we reused our last tea bag, and roasted the meat on a stick over the hot coals. The absence of any side dish or salt did not bother us at all.

Even huge Quiet Lake did its name proud, and without incident we arrived that evening at the place where our adventure had started. It was just five days ago, but felt like a lifetime.

My car contained extra food and cooking utensils, and we intended to sleep under the stars, close to the fire. Just before turning in, we were treated to an out–of–this–world display of *aurora borealis*. Garlands and curtains of light flared up into vivid reds, greens, and yellows, swirling in the grandest celestial show we had ever seen. Our mouths agape, we looked up in ecstasy. As quickly as they had begun, the colours faded again and all that remained were wavering columns of whitish light. For us this spectacular but short–lived celebration was a most fitting farewell to Yukon.

~

Unaware of the fate that awaits us, we tend to pay little heed to the many small, seemingly insignificant and unrelated details of our daily existence that combine to play a pivotal role in our destiny. The Dutch, who are an enterprising but cautious people, have several proverbs describing the unavoidable flow of fateful events: "een ongeluk zit in een klein hoekje," or "kleine voorvallen, grote gevolgen." The pragmatic English equivalent goes: great accidents spring from small causes.

Five years later, when I returned to Yukon in the company of my cheerful wife Irma to make another canoe trip down the Hootalinqua, we ran into a member of the Whitehorse Canoe Club. He told us that the Big Salmon was one the most challenging and dangerous rivers in the area. A dozen members of his club, all experienced river rats, had recently made the same descent Wayne and I embarked on in 1961, from Quiet Lake down to the Yukon bridge on the Dawson highway. None of the canoes had come through without damage, and one of them was lost altogether.

I shudder to think of what had eventually been in store for greenhorns such as Wayne and me if we had not hit the sweeper so early on into the trip. Had we been smart or lucky enough to get by that very first obstacle, we were bound to have come to grief farther down the river, perhaps at a point too far from civilization, making it impossible to struggle our way back out of the wilderness. Having foolishly disregarded all the omens and difficulties which providence had placed in our path to warn us of the impending peril, we had pushed on doggedly like crazy white men until we were stopped cold by the accident that was bound to happen.

In retrospect, hitting the sweeper was a blessing in disguise. Wayne and I easily recovered from that calamity. I for one even appreciated the experience. However, if the same thing had happened much farther down the river, our bones might still be there, scattered by scavengers or nibbled by fish. One thing is for sure, if that had been my preordained destiny, this chapter would never have been written, nor would I have had the privilege of experiencing the adventures described in the following pages.

CHAPTER 4

CANOE COUNTRY IN SASKATCHEWAN

"Come on up!" shouted a voice high above us as we approached the fire lookout tower rising from a clearing in the forest. Looking up at the tall and spindly pylon of steel, we saw that there was a small wooden hut perched on top. A trapdoor in its bottom stood open and the face of an Indian smiled at us. A hand waved us on.

"Going up there? Are you kidding!" Irma exclaimed: "No way!"

"Think of the view!" I offered: "It should be easy. Just don't look down when you climb."

After she relented and we began ascending the first rungs of the long vertical ladder, I repeated my admonition like a mantra: "Don't look down! Just look up!" It was intended to bolster my courage as well as hers. Climbing right below my dear wife, as if to save her from plunging to her death, I kept my eyes nervously glued to her boots. As we gained height, the thrumming of wind in the guy wires sharpened in tone and the entire tower began to vibrate and sway back and forth at a sickening angle.

It was with immense relief that we finally stuck our head inside the trapdoor and hoisted ourselves up above the floor, assisted by the towerman. He was a small, skinny Indian with a perpetual grin on his broad face. His name was Billy.

When we introduced ourselves, he apparently recognized our accent and said: "You guys Dutch? I was over there."

A full–blooded native of the Canadian northwoods visiting Holland? Small world, we wondered. As it turned out, he had served in the army during the second World War. A sharpshooter, he had been engaged in the battle of Arnhem, the locality of some of the fiercest fighting in The Netherlands, aimed at securing a strategic bridge over the Rhine River. Although both my wife and I were quite young during the war, we shared the deep gratitude of all Dutch people toward Canadians and Americans for liberating the Lowlands from Nazi occupation. Billy had been wounded and to show the scar across his belly, he proudly lifted his T–shirt.

A few moments later, looking at his watch, he suddenly interrupted the conversation and pronounced: "Lunch time! See you guys down by the cabin."

Before we could say another word, he had opened the trapdoor and climbed down out of sight, agile as a squirrel.

Seated on a bench, Irma and I looked about us in a very subdued mood. In the centre of the hut stood a table with a large, glass–covered map. There were also a gyroscopic compass, binoculars, and a radio transmitter. Just before leaving us, Billy had explained that he had been up in his tower for most of the day because the fire hazard was extreme. The temperature on this August afternoon was quite warm and the area had not seen a decent rain in a couple of weeks. Moreover, it was very windy, which was exactly the reason why we had left our canoe behind and hiked along the woodland trail to the fire tower.

The hut had windows on all sides and from our precarious perch high above the forest canopy the view was spectacular. All around, trees stretched to the far horizons, swaying in the wind, undulating like the waves of an ocean, an ocean of poplar, spruce, and pine, with here and there the blue of a lake glittering in between. Though impressed by the scenery, we silently realized our predicament.

Here we were, stuck in this isolated eagle's eyrie in Prince Albert National Park, Saskatchewan, in the heart of the great boreal greenbelt that stretched across the top of the globe from Labrador to Alaska and from Siberia to Scandinavia. We were miles away from civilization. But of more immediate concern was the question of how we would ever get down again.

Looking askance at the abyss through the open trapdoor, Irma said with a note of panic in her voice: "I don't want to go!"

The more I pleaded, the more shrill and determined she became: "Just leave me here! You go!"

Finally, to my intense relief, she reluctantly agreed to give it a try. I started first, dangling my shaky legs, feeling for a foothold on the narrow ladder. Then, descending a few steps, I waited for her. It took a near–eternity, or so it seemed, before she gathered the courage to back up to the square hole in the floor and lower her boots. After I guided them one by one onto the metal rungs of the ladder, we began our desperate and silent retreat.

By the time we had reached *terra firma* and walked the short distance downhill to the log cabin, the tea was steaming and the soup ready to eat. Billy was eager to talk. He had been hired by the Department of Forestry as a fire lookout for the summer season. From time to time, during a rainy spell, he was taken out by helicopter for a few days outside, returning with a load of groceries. His employment would end by fall. Soon after freeze–up, he could resume his real calling as a trapper and harness his dogs to the toboggan. Then, he would be alone for long periods too, but that was the way he liked it.

"Do you get many visitors here?" we asked.

"Mostly bears," he said on a matter–of–fact tone. "They can be trouble, breaking into the place."

One night, he had run out and hit a black bear on the head with an axe. Right now, he had a half–decent customer who came every night to collect a fish left for him on a stump. If the beggar was not appeased in this way, Billy feared that the brute might bust a window or break down the door of his wooden shack.

~

The thought of bears reminded us of our own fragile, canvas abode and the canoe which we had left unattended by the shore of Ajawaan Lake, a few miles away. We too had been fishing. No doubt the smell of fried pickerel lingered around the tent, perhaps attracting hungry raiders from far and wide. Concern about bears ruled our daily routine here in the wilderness. The spirit of the feared beast seemed to hover between the trees like a tangible presence that made you look over your shoulder often and spurred you on to more open surroundings where you felt safer.

Saying goodbye to our host, we hurried back along the trail and were relieved to find our lonely camp untouched. The box of food supplies, covered over with a tarp, stood under a shoreline tree a few yards away from the tent and canoe.

Ajawaan Lake was a rather small but pristine body of water surrounded by an imposing wall of boreal forest. Dotted with water lilies and full of fish, the sheltered lake was the former home base of a character who called himself Grey Owl. An immigrant from England, he was employed as a naturalist by the park during the 1930s and played an important role in efforts to reintroduce the beaver. At that time, the aquatic furbearer had become very rare across Canada, after a century of greed by native and non–native trappers. In colonial days, beaver pelts were highly priced on the fashion markets of London and Paris for the manufacture of top hats. By the early 1900s, the bucktoothed rodent, the national mascot of Canada, had been practically wiped out.

A trapper himself, Grey Owl had finally seen the light and stopped killing. He subsequently became a famous author of several books about life in the woods. His British publisher sent him on a promotional tour of England, where his lectures about the lure of the wild charmed a wide audience, including the Queen and members of her royal household.

Grey Owl is still considered one of the first conservationists as well as a colourful charlatan. For instance, he pretended to be a native Indian himself and dressed in buck skin and moccasins. One of his wives was beautiful and intelligent Anaharero, daughter of a native Indian chief, who inspired his concern for the wild denizens of the forest. Living with them was a family of tame beavers, the "little people" as Grey Owl called them. Their log cabin still stands on the shores of Ajawaan Lake. Half of its inside space is taken up by an authentic beaver lodge, a huge mound of logs constructed by the animals themselves. They could come and go as they pleased since a corner of the cabin jutted out over the water. Less well–known is the fact that Grey Owl's love of animals did not extend to the predators. He was ready to shoot any wolf. And when he found tracks of an otter at the outlet of Ajawaan Lake, it too was considered a threat to the beaver family. Traps were set to capture and kill it.

Grey Owl died in 1938 and was buried near the cabin, which has become a shrine in Prince Albert National Park. Its fame has spread, particularly after the 1999 film based on Grey Owl's life. But at the time of our visit, in 1965, we had the place to ourselves.

~

Ajawaan Lake is remote and can only be reached by boat. Today it is possible to drive part of the distance, but during our visit the road ended at the narrows of Waskesiu Lake. From there, as the raven flies, it was about fourteen miles (22 km) to Ajawaan via a huge lake named Kingsmere. In total, it was little more than a day's travel by canoe, but from the start we had been bedeviled by wind, which forced us to take shelter in the woods soon after departure. All we could do for most of that first day was sit in the shade.

The gusty breeze had one positive side: it kept the mosquitoes at bay. By late afternoon, as the wind abated somewhat, we were more than happy to escape from the bugs and launch the canoe again. We soon had our hands full, working our way against the swell close to shore and from point to point. By the time we reached the calmer water on the upwind side, the waves flattened out and soon all our troubles had literally blown over.

To assist with the transport of boats from Waskesiu to Kingsmere, park authorities had provided a light rail link between the two. Loading our canoe and supplies onto the flat rail cart, we pushed it to a point where the stream connecting the two water bodies became deep enough for navigation.

Kingsmere Lake was an imposing sight, some six miles (10 km) long and half as wide. The blustery day all but forgotten, it was great to relax on shore. After sundown, a full moon rose into a luminous sky. Wondering what the morning would bring—perhaps a return of the high winds—we were struck by a bold idea. How about taking advantage of the calm conditions and cross Kingsmere now?

Embarking on the still heaving surface, we set a straight course for the black rim of forest on the opposite side. The vista of water and sky was infinitely wide and peaceful. Sinking into reverie, we were mesmerized by the bright moon and the rhythm of the paddles. Except for their steady gurgle and drip, there was no sound. The heavily laden canoe glided along smoothly over the black, velvety water. The silence was like a balm to the mind, a rare treat these days, even in the wilderness where the throb of engines can be very intrusive. In Grey Owl's time, he and his soul mate must have taken silence for granted.

To us, the one sound that would have been especially welcome now was the cry of the wolf. Grey Owl would probably not have been ecstatic. But he too would have approved of the loon. Its tremulous yodelling that accompanied us on our night–time journey carried far over the languid water and was the universally beloved, haunting music of Canada's canoe country.

~

As we approached our destination on the west end of Kingsmere, we searched for a suitable landing site and came across a wooden boat–launch jutting out from shore. It marked the location of the district warden station. The moonlit roof of the house was just visible between the trees, but there was no sign of life. Suddenly, out of the shadows, a large dog walked up onto the pier and halted a few steps away. It looked exactly like a wolf, a very large one. It gave us quite a start, although it neither barked nor seemed aggressive. As the warden told us later, it was indeed part wolf, part husky. We were to see it again in the next few days, sneaking around our camp like a thief in the night, unapproachable and aloof.

Having studied the map, we knew that the portage trail to Ajawaan Lake was close to the station, a matter of a few hundred yards away. We found it posted with a small, white sign, a welcome beacon in the dark wall of forest. Landing the canoe onto the sandy shore, we were happy to get out and stretch our stiff legs. There was a convenient grassy clearing large enough to place the tent, and before long we crawled into the down bags for a much delayed sleep.

We felt pleased to have made the crossing when we did. All the more so, since the wind picked up again in the morning and turned the lake into a sea of whitecaps. It stayed that way for the next three days.

Beyond Kingsmere lay a chain of smaller lakes threaded together by an ancient Indian trail. The first link was Ajawaan. From there, another portage led to Sanctuary Lake, which was in turn connected to a series of ponds and creeks, extending all the way to the north boundary of Prince Albert National Park. Long ago, this route must have been travelled often by Grey Owl, searching for beaver. His indigenous canoe, constructed from natural materials such as willow branches and birch bark, would have been super light and easy to carry.

By comparison, our craft, though a piece of superb craftsmanship in its own right and made the traditional way from cedar strips and canvas, weighed about ninety pounds (41 kg). It was a little too heavy and unwieldy to be shouldered by one person. Instead, Irma and I carried it together, either at one end, stumbling along the uneven, bumpy portage. The distance to Ajawaan was at least a quarter of a mile (400 m), the day was hot, and in the woods the mosquitoes went wild, attracted by our sweat. After making two additional return trips to ferry all of our supplies across, we made camp by the lake and decided to call it trail's end. This kind of canoe country involved some very hard work indeed, more than we had anticipated!

During the next three days we canoed on Ajawaan, fished for pike and pickerel, and hiked to Billy's fire lookout tower. As to wildlife, we saw little except beavers, which are now common in forested country across Canada. Grey Owl would have been delighted to see that the aquatic engineers had constructed several dams across the creek, raising and maintaining it to a depth sufficient for swimming and diving.

The beavers had also built a huge lodge. Inside, they lived in a vaulted chamber that was above the lake's level. The occupants could come and go unseen through a plunge hole. Now in summer, they lived a life of leisure, but by fall the industrious adults would have to cut a winter food supply of branches and secure them in the deep pool nearby. During this busy time of preparation, they covered the lodge with fresh mud which would freeze into a hard armour to protect their abode from intruders.

Beavers do not hibernate. For the duration of winter, about six months, they live in darkness and subsist on waterlogged bark. Their joy upon emerging from their icy tomb in spring can easily be imagined. They relish fresh greens and care lovingly for their new litter of young, born during winter. By early summer, last year's generation strikes out on its own to search for a mate and find a suitable place to settle down.

In the calm of evening, when "the little people" swam by our camp or sat on shore holding a green twig in their small, monkey–like hands, we could hear their mumbled conversations. According to Grey Owl, beavers form strong bonds with each other and show deep emotions, including sadness upon the loss of a loved one, and loneliness.

With many potential enemies lurking in the woods, the plump and peaceful tribe are ever watchful. Their beady eyes may be poor but their sense of smell is excellent. If one of them becomes aware of danger, the alarm is broadcast in a unique manner. Instantly, the paddle–shaped, muscular tail sweeps upward and slams down onto the surface, producing a sound like a brick thrown into the water.

One evening when we heard repeated warning signals, we quietly rounded a point, hoping to see the cause of the alarm, perhaps a wolf or lynx. It turned out to be a black bear, walking on top of the lodge and ignoring the tail–slapping beavers. The scene reminded us of a story told by Billy. Once he had shot a black bear which showed a large, oval wound in one of its front paws. The unfortunate beast must have broken into an occupied lodge. When its foot groped through the roof, it got a rude reception. There is no doubt that the great yellow incisors of a beaver, capable of chewing through wood, can also serve as a potent weapon!

After our return from Ajawaan, again crossing Kingsmere during the night, we explored another chain of lakes which was then a well–used canoe route, the Lily Lake circuit. It can easily be covered in one day and includes clear, deep ponds as well as marshy shallows where our paddles got stuck in a thick soup of weeds. The interconnecting portages were short but very wet and hazardous. Overgrown with a fragile mat of plants, the boggy ground easily gave way underfoot, threatening to dunk us into a quagmire of mud. Slogging our way through the worst of these muskegs, we had to hang on to the canoe for dear life, struggling to reach water deep enough to float.

Before starting out on our explorations in the park, we had been warned about the danger of muskegs, but we had never expected the bogs to be practically impassable. Today, most of the soft ground on popular trails in the park has doubtless been paved with wooden board walks.

~

Having had a taste of the pleasures and perils of canoe country in the southern edge of the boreal forest, we left Prince Albert Park and drove north to a rocky life zone called the Canadian Shield. Covered with tundra in the far north and with evergreens in the south, the Shield includes more than half of the landmass of Canada. It stretches from the Arctic Ocean down to the northern portion of the prairie provinces and eastward across Ontario and into the United States. Comprised of the earth' oldest precambrian rock, the surface is deeply eroded and plowed over like a gigantic, stony field by the passage of pleistocene glaciers. The landscape is characterized by numerous parallel depressions that are filled with cold, clear lakes. A large–scale map of the region depicts an intriguing mosaic of land and water. It is difficult to say which of the two occupies most space. Since the lakes are often long and narrow and relatively close to each other, they make for ideal canoe country. The portages in between are generally much shorter and on firmer ground than farther south in the woodlands of Prince Albert Park.

The Shield was and still is Indian territory. In the past, aboriginal families travelled widely hunting for game that is thinly spaced in this vast and unproductive habitat. Though inhospitable, the immense, empty spaces beckon and compel one to go ever onward to the next bend, over the next ridge, to look for whatever it is that one seeks.

The Indian way of life, of covering mile upon mile by birch bark canoe, was copied and exploited to its full potential by the first European scouts who ventured westward across Canada in the late 1700s. Their explorations had commercial goals, to make contact with the natives and induce them to kill furbearing mammals. French *voyageurs du bois* strove to stay ahead of the competing Scotsmen of the Hudson's Bay Company, who were not far behind and soon dominated all trade, bartering fur for traps, rifles, axes, pots and pans. Beaver skins were the currency by which the value of all trade goods was measured and compared.

Canoe travel on the Shield is a tradition that has survived to this day, particularly in Ontario and in the adjacent States. The romantic lure of "the singing forest" and "the song of the paddle" inspires nature lovers of all ages. In the spirit of the hardy voyageurs, the general idea is to visit the most lakes in the least amount of time. However, Irma and I were of a different bend. Instead of seeking to maximize our mileage, we preferred to get the most out of a limited number of localities.

~

Starting near Lac La Ronge, we canoed north to the Churchill River system via an interconnected series of lakes, always checking our progress on the large–scale map. Here, in this confusing labyrinth of deeply indented bays and scattered islands, it was easy to lose one's way. Fortunately, portages were well–trodden by other travellers, of which we saw very few. Less fortunate was the fact that we were again impeded by high winds as well as rain. More than once, we ended up weather–bound on some rocky shore. On other days, the country was enveloped in the smoke of forest fires drifting down from far away. In the vast northern wilderness, hundreds of conflagrations can be burning unchecked, started by thunderstorms, a near everyday occurrence during spells of hot weather.

One night, a storm unleashed its full fury overhead. The awesome, primordial forces shook the ground under our tent. Or was it just our imagination? Holding onto each other and trembling like scared rabbits, we felt ready to crawl into a cave, cowering under the elemental and eternal energy of this wild land.

Before long, we ran short of supplies, although we had often augmented our larder with blue berries and fish. Pike and pickerel were there for the taking until our last lure was lost. Canoeing on an empty stomach was no fun at all.

As to wildlife, we saw a few waterfowl, bald eagles, the odd muskrat or beaver. At this latitude, there were neither deer nor elk. Except for a rare caribou, hidden in the vast void, the only common cervid in the northern Shield is the moose. They too are few and far between in this stony habitat and harsh climate. Their population is further depressed to minimum levels by hunting, superimposed on natural predation by wolves and bears.

The average density of moose within Prince Albert Park has been estimated at one animal per square mile (2.5 km^2). By comparison, in the Shield, on less fertile soils and without year–round protection from the gun, moose number perhaps less than one per ten square miles (25 km^2)! Looking for them is like searching for the proverbial needle in a haystack.

Small wonder we spotted no large mammals at all, particularly since we travelled routes that were regularly used by hunters. Moreover, even if we had unwittingly come close to any animals, they would have been hidden from view by dense shoreline vegetation. We did not even find tracks. The shores were mostly rocky, the few beaches narrow and wave–washed. We neither found the heart–shaped hooves of moose, nor the clawed pads of wolf or bear.

In summing up, in our experience, the obvious merits of canoe country did not lie in its opportunities for watching animals, but in its virgin spaces, clear waters, and peaceful solitude.

PART II

ROCKY MOUNTAIN WANDERER

Chapter 5

Bivouac, Baits and Betrayal

"Foothill Forests Overrun With Wolves!" ran a sensational headline in a January 1961 issue of the *Calgary Herald*. The article stated that a pilot flying for an oil exploration company had spotted a pack of wolves near the Brazeau River west of Rocky Mountain House, and that he had seen the remains of several of their kills. Reacting to the story, the local hunting association, fearing that the country was being overrun by the predators, had already demanded that control measures be taken by government agencies.

The press report sounded very exaggerated and contrary to my own experience. My searches for wolves in the Rocky Mountains and foothills, coupled with inquiries from forestry rangers and park wardens had turned up nothing. In those days, wolves were definitely scarce in all of western Alberta, and the reason was no mystery at all. During the preceding decade, the province had unleashed the most vicious poisoning campaign ever in Canada and perhaps the entire world. Its objective had been to wipe out all wild canids in settled regions so as to combat the spread of rabies, which had first reared its ugly head in the North. Between 1952 and 1956, a staggering amount of poison, close to one million units, was distributed free of charge to trappers and landowners. The toll they took would have been utterly unbelievable had it not been monitored by the government's own agents. Over four winters, the total kill included 5,200 wolves, 171,000 coyotes, and 55,000 foxes! Plus untold numbers of other scavengers, birds as well as mammals, for which no data are available.

So how could it be that wolves were suddenly supposed to be numerous? Granted, it was not impossible that the pilot, from his high vantage point, had indeed observed a pack on the frozen, snow–covered river. The very thought conjured up exciting images and gave rise to a bold idea: How about going there to look for myself? After discussing the matter with my friend Bill, an experienced outdoorsman, we decided to leave the following weekend. Our first stop was the forestry ranger station in Rocky Mountain House.

Wildlife officer William was incredulous and unaware of the *Calgary Herald's* article: "Wolves overrunning this country? I haven't seen a track in years!" They used to be common in his district though. During the fifties, he had personally poisoned sixty–seven of them in a single winter! "Ah, if it hadn't been for rabies, I would not have bothered. I got nothing against wolves. You kidding? I like them better than people!"

This seemed a strange statement to make. However, Bill, always looking for the best in his fellow man, thought that William was an okay guy and squarely on side of the much persecuted wolf. As we left his office, the ranger made a last–minute request: "Let me know what you guys find."

It was getting close to dusk when we headed out of town and turned off on the bush road leading into the direction of the Brazeau River. The narrow track, used mainly by oil and forestry company vehicles, had recently been cleared of snow which was piled high along the shoulders. The surface was icy, forcing us several times to back down gingerly and take another run at a slippery hill. There was no one else on the road. The car's headlight illuminated nothing but trees and more trees. Snow flurries made our prospects look all the more bleak and lonely. Moreover, we were unsure as to the distance we had to drive. Before starting out, we had studied a large–scale map and marked a point where the road appeared to come closest to the Brazeau River. This is where we intended to begin our two–day cross–country trip tomorrow morning. However, we had no idea where we were going to spend the coming night.

Our prospects began to look up after we crested the next hill. The snow–streaked sky above was aglow with the bright lights of an oil drilling rig, rising up from the surrounding forest like a gigantic Christmas tree.

"Let's stop here," suggested Bill: "And ask these guys if they have seen any wolves."

The burly foreman was in his trailer and he had plenty of time for a chat: "You guys looking for wolves? Ah, there's nothing here. You want to go north, to the Peace River country. I was working there last winter and we saw lots of wolves. They like to travel plowed roads, you know, to get away from the deep snow in the forest. We run them down every chance we get. They're terribly vicious, these buggers. One of them still tried to bite me while he was lying on the road—his spine crushed by the truck—and I had stopped to take him out of his misery...."

I was disgusted, but not surprised. At that time, in my dealings with hunters and trappers I had come across so much callousness and hatred toward wolves, that, by comparison, the predators began to look like angels. Fortunately, easy–going Bill did all the talking. Ignoring what he did nor care to hear, he chatted about the newspaper report that wolves were overrunning the country west of here, and that we would be heading out to the Brazeau River tomorrow to investigate. "You wouldn't have a place for us to stay here overnight, would you?" he threw in casually.

"Sure thing!" the foreman responded brightly and he gave us the number of one of the bunkhouse trailers we could use. It was warm and comfortable. What a relief to get out of the cold and dark!

Next morning, rising well before it was light outside, we joined some of the crew for a copious breakfast, all we could eat, courtesy of the company! The foreman, who had been on call all night, was asleep, and without saying goodbye to our generous host, we left the noise and bustle of the drilling rig behind, eager to continue our single–minded quest.

It was quite cold, but the sky was clear and there was an inch (2.5 cm) or so of freshly fallen snow on the ground, ideal for tracking. After a few miles, we reached an intersection where an unplowed cutline angled off roughly into the direction we wanted to continue on foot. After parking the car on the side of the road, we loaded our camping gear onto a flat toboggan, and, dragging it behind us, set out through the snow that reached halfway to our knees. "It'll be much easier once we get to the river," Bill said.

The cutline ended somewhere in the middle of the bush and our progress became even slower until we finally saw the trees end. Sliding down the river bank, our hearts leapt at the sun–lit scenery that opened up before us, awesome yet inviting, stretching as far as the eye could see to the Rocky Mountains glittering on the western skyline. What immense spaces! Plenty of room here for wolves, it seemed. To see them might be too much to wish for, but we had high hopes that we would at least hear them howl or find their tracks.

On the river, the going was indeed far better than it had been in the forest. The snow was less deep and wind–packed. In our youthful ignorance, we never considered the risk of breaking through thin ice. In fact, staying well away from the drifts along shore, we often got close to leads of open water where we could hear the current gurgling under our feet.

The afternoon glided by and we proceeded in high spirits, mile after mile. The weather had turned out ideal: calm and not too cold. As to animals, we saw nothing, not even a fresh track. The snow was as immaculate as a sheet of blank paper. In this white and virgin void, we suddenly heard the clanging of a bell coming from the woods around the next bend of the river. A church? Bill, who had worked in the forest in the past, knew what it meant: "That's a dinner bell! There must be a lumber camp over there!"

Hidden in the trees not far from the river bank, we discovered a collection of huts and large tents. The saws were idle for it was lunch time. The workers, about a dozen hardy–looking types, were sitting at a long table in the cook's shack. They did not talk much and hardly blinked an eye at our sudden entry from nowhere.

Bill, his usual smiling self, intended to inquire about wolves. Before entering camp, he had casually advised me that we should pretend to be hunters. "If you tell these guys that we're just looking for wildlife, they might think we're some sort of sissies."

"Hunting wolves?" one of the guys responded curtly: "We seen nothing."

"You hungry?" interjected the cook: "Find yourself a seat."

The meal consisted of lots of meat and gravy, potatoes and beans, as well as plenty of sweet desert. Never in my life have I seen people wolf down that much at one sitting! These guys had appetites! As a matter of fact, so had we.

Richly satiated, it was great to be on our way again in the splendour of the wintry wilderness.

There were just a few hours left before sundown and we soon began to look for a suitable spot to spend the night, which proved surprisingly difficult. The tall trees on the banks were mostly poplars that afforded a minimum of shelter. But once we settled on a site, the preparations needed for a bivouac were simple since we did not have a tent. We just scraped the snow away, piled a number of spruce branches on the frozen ground and spread our sleeping bags. A huge fire was quickly made and soon our billy can was coming to a boil.

The sky remained luminous until well after sundown, and we sat by the glowing embers until stars began to twinkle. Crawling into the sleeping bags, we kept all our clothes on, except the jackets which we draped over our shoulders and hips. Awakening briefly, I saw that a thin arc of moon had risen, while the temperature was obviously dropping. The frigid air stung my face as I listened to the sub–arctic night. Far off, an owl called. From time to time, the river ice rumbled like distant thunder, and the bark of trees shrunk and cracked with the sound of rifle fire. But the sound that I wanted to hear most of all, did not come.

The morning dawned clear and bitterly cold, but the absence of wind made it bearable. While I was building a blazing fire, Bill was greatly amused to find that his slab of bacon had frozen rock–hard. The only way to cut it up into manageable pieces was by using his big axe.

Breakfast over, our plans were simple: go back the same way we had come. At least, the going was easier for we could drag the toboggan over the same track as yesterday's. By the time we reached the road, the temperature had risen a bit and the old Pontiac, to our relief, started readily.

We had not yet given up hope of finding wolf sign, and on the way back to Rocky Mountain House, we kept a sharp eye on the ground ahead. Two days had passed since the last snowfall and the bush road had not been cleared in the interval. The hoof marks of moose, elk, and deer were quite common. Here and there, a neat string of closely spaced prints turned out to be those of lynx. But mile after mile passed without any tracks of wolves. We were almost ready to give up on our time–consuming stop–and–go routine, when we noticed yellow urine stains against the snowbank on the road shoulder. Coyotes again? But look at the size of those foot prints!

I was greatly excited. These were indeed wolf tracks, the very first I had ever found in snow! We followed them for a ways down the road to where the animals had turned off down a creek bed leading into an open area of muskeg. It looked like there had been four or five in the pack. It was too late in the day to track them farther. We had to get back to the city, but we promised ourselves that we would be back here again next week. With luck, the tracks might still be legible. Like an open book, they would allow us to enter into the secret world of *Canis lupus*, the legendary outlaw who had managed to survive here in these forests, despite all the poisoning of the recent past.

"We actually should stop in Rocky and tell the ranger what we found," said Bill: "Otherwise, he might think we are just a couple of amateurs."

As the driver, I disagreed. It was late and we had still more than two hundred miles (300 km) of highway ahead of us to get back to the city. "Maybe next time," I said.

~

One week later, on our way in through Rocky Mountain House, we stopped by William's office and reported the highlights of our previous trip. When we described the location of the tracks, he seemed to know exactly which creek and muskeg we were talking about. He agreed with us that it was indeed nice to know that there were a few wolves left in the country.

"Thanks for telling me," William said: "Incidentally, a couple of days ago, a pest control crew was here, from the Department of Agriculture. They got the job of wolf control here, even though this is not their zone but a forestry district."

"Oh, well," he shrugged: "Nothing I can do about it. That's politics for you."

When we asked him where the poison baits were going to be set out, William said that the crew had intended to charter an aircraft or helicopter to take them to the headwaters of the Brazeau River, not far from the eastern boundary of Jasper National Park. He seemed just as sorry about this nasty business as we were, and the idea that he might betray our information to the poisoners never occurred to us.

Any depressing thoughts soon lifted from our minds as we drove down the bush road. There had been another fall of fresh snow, obliterating all sign of last week. However, to our delight, upon arriving near the muskeg, we again found very fresh spoor! Like last week, tracking conditions were perfect. The weather was less cold than before, and we had all day ahead of us.

Beyond the muskeg was a snow–covered lake where the pack had fanned out, allowing us to establish that there were indeed five animals in the pack. In the middle of the frozen expanse, the animals had rested and played, running about and chasing each other like young dogs. Several of these wolves might be pups of last summer.

On leaving the lake, the pack had come together again in single file and ascended a forested hillside via an easy ridge–top route that led to another muskeg. Here in the open, the animals had spread out once more. At one point they bounded along as if they had sighted and chased prey. They proved to be on the trail of a moose, also running. Very soon the overall picture became quite confusing and we lost track of most of the wolves. Eventually, circling the muskeg, we picked up the exit trail of the pack. There was no evidence that the chase had been successful.

As the afternoon progressed, a Chinook wind sprung up and the temperature rose well above freezing. The snow became wet and slushy. Shaking in the breeze, the snow–laden branches dropped their load, obliterating wolf sign below. Eventually, we again lost track of most of the pack and ended up trailing a single animal. It led us to several old dens! Two were no more than single burrows in a hillside, but the last one had been excavated under the roots of a large jack pine by the shore of a small lake. The wolf, likely a female, had done some spring cleaning here, scraping out the accumulated debris and snow. Long, grey hairs were clinging to the roof of the den's entrance. It looked like she was hunting for a home for her future progeny, now stirring in her womb.

Dreaming about our prospects for wolf watching here the coming summer, we sat down for a rest before heading back to the road.

~

Bill could not make another trip for the next few weeks, and I returned alone, hoping to do some more tracking. The weather had stayed warm and there was not much snow left on the ground. Stopping at the same place as before, I got a nasty surprise. Tacked to a prominent tree was a yellow warning sign: "Poison baits!" In disbelief, I read on: "Keep dogs away! In the interest of protecting poultry, livestock, and game...."

Shocked and depressed, I walked down the muskeg to the lake where the pack had rested and played a few weeks earlier. There were no tracks now, but on the shore, wired to a metal stake in the frozen ground, lay the bait. It consisted of the head and neck of a horse. The toxin used was evidently the dreaded ten–eighty, or sodium fluoro–acetate. It is injected into the meat of a living horse and, after its death, permeates every bit of the meat.

Later, when I told Bill, he exclaimed: "That guy William is a rat!" Unfortunately, confronting our false friend would be useless now. Instead, we decided to write the government department responsible for the poisoning, not to complain about its two–faced wildlife officer, but to protest the placing of baits in an area where we felt big game animals were plentiful and wolves few in number.

Realizing that a letter from an individual would mean less to those in power than the concerted voice of a group, Bill and I started The Alberta Wolf Listeners Society. It was based on an idea obtained from a magazine article about like–minded souls in Ontario, who slept under the stars in Algonquin Provincial Park for the special purpose of hearing wolf song. At that time, the Ontario Wolf Listeners Society had no more members than our own newly established group. Of course, in subsequent years, the idea caught on, and today, in the eyes of millions of city dwellers, the formerly feared varmint is almost as popular as Bambi. In Algonquin Park, thousands of tourists eagerly join wolf–howling excursions led by naturalists each summer.

This new and positive angle to the wolf question, the fact that a growing number of people no longer hated or feared wolves, nor resented their depredations on wild prey, became our main point in a letter to the provincial Director of Forestry. We also argued that wolves were too few in number to constitute a serious threat to game populations or to livestock, and that poison baits were bound to result in the needless destruction of other furbearing mammals as well as scavenging birds. We also asked that proper research into the wolf's impact on the ecosystem be conducted before any further control was to be undertaken.

In his reply, the Director explained that he had delayed answering our letter until the baits had been picked up. He could now state that "there is very little evidence of any animals having taken meat from them." And he added that his office "had received a number of complaints to the effect that bands of wolves were found in the area northwest of Nordegg, and received through the mail a clipping from the *Calgary Herald* on the same subject. It was therefore decided to place some baits to determine the results."

In other words, poison first, research later. I saved that letter as an illustration of the way in which predator control was practiced in Alberta in the bad old days, until better times arrived and wolves staged a strong comeback.

CHAPTER 6

WOLF VALLEY

"Ijust saw a wolf!" Cedric exclaimed breathlessly as he came running back to camp: "On the meadow! No kidding!"

His excitement was understandable. Not only was it the first wolf he had ever seen in his life, but it happened on the last morning of a three–day hiking trip intended specifically to find wolves, and thusfar we had neither seen nor heard any.

Rare wildlife sightings are all the more rewarding if you can share them with a companion. However, if only one of you is lucky, the joy is marred by the other's disappointment. There are those who are so sensitive of other people's feelings that they, upon asked, would rather deny having seen anything at all. "You didn't miss a thing," is bound to bring a relieved smile on the other's face. But I for one prefer to hear the truth even though it might give me a pang of envy. Sightings by others complement and enrich your own experience, providing you can rely on the accuracy of their report. Not everyone is a good observer or pays more than casual attention to the same favourite species as you. Cedric and I had done a lot of birdwatching together and I knew that he was not given to exaggeration or jumping to false conclusions. Yet, I could not help grilling him a little since he had seen no more than a brief glimpse of the animal running into the trees.

"Are you sure it wasn't a coyote?"

"No way! This thing was huge!"

~

That morning we had crawled out of our tents early. We had fourteen miles (23 km) of hiking ahead of us to reach our car, which was parked at Rock Lake. Clear skies were promising a hot day. So, we had decided to leave soonest. First to finish breakfast and packing up, Cedric had strolled over to the mountain meadow to take a last look at the deer and elk we had watched there every morning and evening. A few minutes later he had come running back into camp with his exciting news.

"Let's go back. Maybe there are other wolves to follow."

Silently, we hurried through a narrow belt of trees to the edge of the natural meadow, which was about half a mile across and studded with low willows and dwarf birch. Raising the binoculars, we carefully scanned the grassy openings but saw nothing. No deer, no elk, and no wolves.

Hoping that the animal spotted by Cedric might still be within earshot, I imitated a wolf howl. A minute later, something stirred on the opposite side of the meadow. Through the glasses, we saw four wolves file out of the trees. Two were light grey, one was black, and the fourth a silvery colour. The animals halted, looking at us. Then, two of them, a grey and the black, trotted into our direction. We sat down on a handy log and watched spellbound.

While the silvery wolf stayed behind and lay down, one of the grey animals approached in a roundabout way, skirting the edge of the meadow that became narrower toward our end. When the wolf was directly across from us, it studied us briefly, then moved off into the trees, leaving us with the uneasy thought that it might circle through the woods and sneak up on us from behind. In the meantime, the other two wolves had paused on a slight rise near the centre of the meadow. They just stood there, a little less than a hundred yards (90 m) away, giving us an excellent opportunity to study them.

The black wolf had a grizzled muzzle. Through the glasses, we could see that its eyes were a pale gold, their expression harmless as a child's. Its grey companion was about a third bigger and had a stocky, powerful body. His colours ranged from creamy–white on undersides and legs, to dark grey along the spine, and chestnut on the tail, which was tipped with black. His face was whitish with a dark mask over the eyes, their gaze inscrutable. Both wolves glanced at us furtively as if we had ceased to be of interest. Perhaps they were mystified by our immobility. We neither moved, nor made a sound, mesmerized by the magic of this meeting.

Meanwhile, the sun had climbed over the mountainous skyline to the east and long beams of light crept out over the uneven surface of the meadow, setting a million stars twinkling on the hoarfrost that coated every sprig and blade of grass. When a ray of sun touched the two wolves, they stood out brilliantly against the deep shadows of the woods, like jewels on velvet. Beyond, golden spires of spruces reached into fathomless blue.

To our delight, when a faraway howl echoed over the hills, the two animals on the meadow pointed their noses upward at a slight angle and responded with a duet. Their fangs gleamed in their half open mouths, and their breath steamed in the frosty air. The black wolf had a higher voice than the big grey. The silvery animal, which had stayed behind, also had a high–pitched voice, particularly as compared to the distant wolf, which we could not see. In the meantime, the second grey wolf, the one that had gone into the woods, had reappeared and joined the pair in the centre of the meadow. The three of them howled at intervals.

Presently the trio fell silent and there was no further response from afar. Staring intently at us, one of the grey wolves suddenly seemed to have made up his mind and started to come directly toward us. Uncertain as to where his menacing approach would end, we could not help beginning to feel uneasy. But before the situation became critical, a hoarse call summoned everyone's attention to the other side of the meadow. There, just visible among the trees, stood a great lanky wolf, dark grey with black head and legs. He had come to get his straying companions.

All at once, the three wolves lost interest in us and turned away to cross the meadow. Rejoining the silvery animal, they fell in line with their black leader and trotted off into the woods, leaving us with nothing but a shared memory.

For me, unlike my friend, that bewitching encounter did not represent my first wolf sighting, but it was only the second one in Jasper National Park, after years of hiking, and to this day it remains one of the most enchanting. The mountain meadow turned out to be a key location, a rendezvous site, where the local pack brought its pups after they leave the natal den. Subsequently, I went there each year over three decades, sometimes as often as two or three times a month from June to October.

Chapter 7

Hiking the Horse Trail

The Willow Creek Trail begins near Rock Lake just north of the park. After one mile or so, it turns off from the Willmore Wilderness Trail and follows a wooded ridge scarred by forest fires. The boundary of Jasper Park is marked by a simple sign and beyond there, the trail descends to Rock Creek and continues on the other side to Willow Creek. In between these two pristine streams, separated by a few hundred yards of practically level forest, lies the lowest watershed divide you may ever see. Rock Creek drains north, while Willow Creek flows the other way to join the Snake Indian River, which drains much of Jasper Park's vast and virgin northern hinterland of forests and mountains. The valley bottom near the confluence of the Snake Indian River and Willow Creek is characterized by a mosaic of large and small montane meadows. In this semi–open terrain, the hiker has a fair chance of spotting wildlife. Deer and elk are quite common, as are bears and wolves.

In 1965 when Irma and I made our first back–packing exploration into the valley, the trail was so little used that it was easily lost, especially on the braided gravel bars of mercurial Rock Creek. Not only was its wandering main channel always a challenge to ford, but the shoreline woods could be inundated by sudden floods, obscuring all signs of a path. Moreover, in those days, the official Jasper Park map was highly misleading. It indicated that the trail continued south of the creek by turning off sharply and ascending the steep Bosche Ridge, instead of just following the valley bottom. This seemed illogical, so Irma and I decided to proceed by bushwhacking. As luck would have it, we soon stumbled on the real trail that led into a direct line to the district warden station, a few miles down.

However, before we got there, our path was intersected by three other creeks, or so we thought. In fact, as we found out later, they were meanders of Willow Creek, a crystal–clear, knee–deep watercourse. By leaving the trail and simply following the banks for a little ways to the third ford, we could have saved ourselves the trouble of taking off our boots twice! Horse–mounted travellers, such as park wardens and outfitters, then just about the only people to venture this far, approach creek crossings quite differently than pedestrians do. Riders just follow the shortest route. Hikers are happiest if they can keep their feet dry. Some people have what it takes to wade barefoot through mountain streams, but not us.

Rock Creek can be especially tough to ford. Unable to cope with the numbing cold and the uneven, stony bottom, Irma and I had kept our boots on. Drying them later was another matter. During subsequent trips, just for the purpose of fording the main channel, I brought sneakers. Conscious of the accumulating weight in my pack and paring it down to the essentials, I carried the extra footwear only as far as needed and stashed them near the bank, hoping they would still be there next time. Unfortunately, this was not always the case....

One of the advantages of having company on the Willow Creek hike was that one could share sneakers, saving one guy the bother of packing his own. The first one to cross Rock Creek threw them back to his partner. One day, I was accompanied by Otto, a British–educated physician. The doctor was an experienced naturalist and he always hiked on rubber boots, the best and cheapest way to keep his feet dry in these rain–soaked mountains. Otto was nice enough to share his footwear with me for the specific purpose of getting across Rock Creek. After wading to the other side of the channel, which was about fifteen yards (14 m) wide, he took his boots off, drained the water out of them, and threw the first one across. It went straight up in the air and plunged down into the middle of the raging torrent, while I was waiting barefoot on the other side.

SNAKE INDIAN VALLEY
Jasper National Park

SCALE 0   6 Miles / 10 km

"Sorry, old cock!" was all Otto said in his British accent. He then just stood there, without making any effort to recover his fast–disappearing footwear. I, the most thin–skinned tenderfoot you ever saw, was forced to run on the gravel alongside the creek to overtake his bobbing boot and retrieve it from the icy–cold water.

I hold no grudge against Otto nor blame him for his clumsy toss, for I have done the very same thing twice to other companions, including my brother Fred visiting from overseas! Throwing a boot or shoe over a wide, turbulent creek is intimidating. One has a tendency to hold on just a split–second too long, so that the object goes straight up instead of forward. With me as the culprit, both my companion and I ran on either side of the creek until I gave up, yelping and cussing, to let the other guy make the retrieval.

So much for the challenging old days! Today, the hiker will find Rock Creek bridged, at least the deepest channel. However, after heavy rain the swollen torrent may switch over to the other side of its wide, gravel bed and leave the bridge high and dry. Or the flimsy structure may collapse after the stream undermines its foundation.

The increasing number of tourists that visit the Willow Creek district have led to several negative consequences for hikers using the trail. The once so pleasant woodland path that gives access from the north, on which Irma and I found the tracks of wolf and bear, is now worn down by horses and washed out by runoff. It has turned into an ankle–spraining obstacle course of stones and boulders. Inside the park, on level ground through the scenic heart of the Willow Creek meadows, the trail has become so deeply rutted in wet spots that the wardens found it necessary to reroute it to the adjacent woods. It now alternately climbs and descends across the contour of the hillside. Not only is this section of trail tough on tired backpackers, but horse traffic has done serious damage to the fragile, spring–veined soil at the base of the slope. Once covered in deep moss, some depressions have been trampled into a deep mire of mud so that trail maintenance crews had to do extensive damage control including the construction of bridges and boardwalks.

The park's increasing popularity among hikers and riders has also taken some of the fun out of overnight camping. Prior to the mid 1970s, it was possible to place your tent in any pretty or convenient location. If you were overtaken by fatigue or worried about an impending change in the weather, you could stop and make a fire anywhere you fancied. The hiker of this nostalgic era, seeking the simple charms of a secluded wilderness camp and doing only what was then perfectly acceptable, now faces court appearances and fines like a common criminal. Stringent regulations not only limit the number of travellers allowed on the trail on a day–to–day basis, but camping and fires are restricted to designated sites, which are about eight miles (12 km) or so apart. The ground on the most popular sites soon becomes bare, and the trees fail to provide much shelter from wind and rain after most of their lower branches are stripped for fire wood. To the tired and wet backpacker, a muddy and drafty campsite is the last thing he or she needs for an enjoyable backcountry experience. Moreover, the most frequently used sites become magnets for scavenging bears.

The reasons for the changed regulations are understandable, an unavoidable consequence of increased visitation. You never know, however, how many people will stay overnight at the same campsite as you. On some days you can still have the place to yourself. On others, your peace and quiet may be shattered by the arrival of an outfitter with a group of riders and a pack train of two dozen horses. The wranglers and guides have much work to do, such as cutting fire wood with noisy power saws, and pounding in stakes for the group tents, which require spacious and level ground. Sometimes hikers are asked or told to move out of the way. In one instance, four guys and I, annoyed by the intrusion of a large horse party, pulled up our pup tents and decided to go elsewhere. To avoid friction, horse parties and backpackers are now segregated. The hikers were the losers since they were directed to camp on the other side of the stream, where the woods are more open to the northwind and offer even less protection than the heavily–worn old site.

On some trails, such as those leading to popular fishing spots, there have been nasty arguments or worse between two–legged travellers and those astride the four–legged kind. The hiker's point of view is quite obvious; horses can ruin the trail and leave manure all over, occasionally even on camp sites where they are not supposed to be. By contrast, the outfitter's objections to the growing hordes of backpackers are more subtle. Of course, they resent it if hikers raid their stack of fire wood and burn up tent poles left behind for the next trip. However, their unspoken enmity goes deeper and has historic overtones. Fact is that some outfits go back a very long time, long before hiking became popular. When Jasper National Park was established in 1907, the Canadian government encouraged outfitters to operate in the park and provide services for tourists. All the major backcountry trails were blazed by people on horses. Some follow ancient routes that go back to Indian days. Most traditional campsites such as Willow Creek have been used by outfitters for two or three generations. No wonder they resent being told that they are no longer welcome and needed.

When I began my search for wolves in Jasper, the horse parties consisted mainly of anglers and big game hunters. On their way to the high country north of the park, large pack trains routinely travelled through the Snake Indian Valley. These outfits were even permitted to overwinter their horses in the meadows, leaving the animals to fend for themselves until they were rounded up for the short summer and fall business season. Up until the early 1970s it was common to encounter herds of two or three dozen horses on the rendezvous site where Cedric and I had seen the wolves. The sound of their bells was a frequent annoyance, especially at night.

The once unquestioned grazing rights for private enterprises were terminated in the mid 1970s, when it was no longer deemed appropriate that horses were competing for forage with indigenous herbivores such as the elk. The outfitters now winter their stock outside the park, while their summer use of the backcountry has been curtailed to the benefit of the hiking community.

However, even this seemingly simple issue has a flip side. The reduction of horse travel has led to the obliteration of trail sections which have become grown over or obscured by vegetation, making them hard to locate for hikers and increasing their risk of getting lost in unfamiliar country. Secondary trails have faded away altogether, such as a convenient short-cut between the upper and lower Snake Indian trails that I often took in the past. Here, even the wolves should have a complaint to air, since this route, like all game trails and human pathways in the mountains, was once a regular highway for the local pack! For me, the mud and dust on horse trails was like a news bulletin that I keenly examined for the signatures of the elusive critters of the forest. On most days, all I got to see of them were their tracks superimposed on equine hooves.

~

The absence of horses has had implications for another, more serious problem that has been accelerating of late, namely the changing vegetation of the meadows themselves. Nearly everywhere, the grasslands are being invaded by bushes and trees. This is a natural succession of plant communities. The climax ground cover of this valley is a closed forest of predominantly spruce and pine. These hardy conifers are the dominant tree species that have been advancing into the foothills and mountains of the West ever since the last continental ice–sheet retreated, some ten thousand years ago.

We indeed have to go far back in time to fully understand the primary dynamics of this landscape, which is very young by geological standards. In the wake of the retreating glaciers, the gouged–out valleys became water–filled. As the shallower lakes and marshes gradually drained away or evaporated, the first plants to colonize the virgin soil were sedges and grasses. Like the foothills and prairies to the east, the Willow Creek Valley was once an open steppe, grazed by bison and other herbivores. The hoofed mammals in turn attracted wolves, as well as two–legged hunters.

The nomadic tribes of North America thrived on the open plains and in the wide mountain valleys, which allowed ease of travel and relative safety from enemies that could be spotted from afar. After the grass steppe evolved into a savanna with scattered groves of trees, the stone–age savages were smart enough to do something about it and advance their own interests with a simple management technique. To prevent the trees from closing ranks, the Indians put them to the torch. They knew that the bison would return soon after the scorched earth had turned green again with vigorous young grasses.

The deliberate incineration of forests in western Canada lasted until the end of the nineteenth century. A turning point was the establishment of national parks. One of their key management philosophies became fire prevention. Since then, much of Jasper's formerly open and semi–open terrain on the mountain sides and in the bottomlands has become carpeted with trees. Early photographs taken from viewpoints in the main Athabasca Valley show a striking difference with today's picture. The process of forest encroachment on grassland is ongoing everywhere in the lower montane, particularly in the Willow Creek district after the once heavy grazing pressure of horses and other herbivores had greatly diminished.

~

Intertwined with the history of the horses is the story of the local elk. By the time Irma and I first visited the Willow Creek Valley, the population included some two hundred head. In a way, these grazers were maintaining their preferred habitat, that of the open grasslands, for they consumed sprouting bushes and trees. However, their impact on aspen poplar was locally severe because this deciduous tree only occurs on a few sunny, open locations in the lower montane. In fact, during the 1960s, the aspen groves in the Willow Creek district were a dying stand as all young shoots were removed by the elk and the horses. Blow–downs littered the ground.

The demise of the poplar was a calamity that reverberated down the food chain. Beavers, which had once been common, had died off or been forced to move out. Their deserted lodges and dams, surrounded by the stumps of felled aspen, still stand in the creek as monuments to their industry, which had benefitted a host of other life forms as well, including fish, birds, and insects.

The situation began an abrupt reversal after 1974, when the elk population nose–dived, precipitated by an extremely severe winter. Deep snow and overgrazed range led to massive starvation and poor reproduction among the elk, while their predators, the wolves, had returned with a vengeance. Through the two–pronged impact of starvation and predation, the park's elk population dropped to about one quarter of peak numbers. In Willow Creek the local herd levelled off at no more than a dozen or two.

The reduced elk presence, which coincided with the banishment of the horses, contributed to a dramatic acceleration in plant succession, which was no longer retarded by grazing. Inexorably, willow and bog birch began to choke grassy terrain. Tiny spruce seedlings that managed to sprout despite competition from the grass were no longer nibbled or trampled. Poplar groves expanded with new vigour by sending out suckering shoots that would never have had a chance to mature prior to the 1970s. A decade later, dense stands of saplings replaced their dying predecessors, hiding the deadfall from sight. On nearly all open locations, poplars were gaining ground and so were the willows.

~

The growing loss of the grasslands was viewed with concern by Parks Canada officials. Finally, after much discussion, the warden service decided to try to reverse the process by using the same ancient tool as the aboriginal tribes of old: fire. Did this mean that humans were going to interfere with nature, here in this sacred sanctuary? Yes and no, depending on one's point of view. The park service's argument was that fires had been part of this ecosystem for centuries. Soil samples, which showed variously aged strata of ash and charcoal, proved that fire had been the catalyst of a long–term cycle of forest rejuvenation. This dynamic process had been interrupted by the park's policy of fire suppression. The question remains, however, whether the conflagrations of the past had been ignited by mankind or sparked by nature's own arsonist: lightning.

In a complete reversal of long–standing policy, park managers argued that prescribed burns were a practical and even natural way of restoring the few remaining grasslands. A prime candidate was the Willow Creek Valley. However, quite different from the Indian way of doing things, which involved no more than letting the flames go with the wind, the wardens crippled their own efforts by wanting to raze only the meadows and not the adjacent hillsides. This was a tricky business, requiring a lot of strategy, manpower, and luck. The upshot was that only part of the meadows burned. Even repeating the process the following year proved only partially successful. Much of the gains made were soon lost to the irrepressible willow roots. The charred skeletons of the bushes began to green up the very next summer.

Those of you who, after reading this chapter, decide to see the valley for yourselves will have the last word as to what the place looks like. I have not been there for a few years. At the very least, I hope to have given you a perspective of what once was. In 1965 when Irma and I hiked the trail through the meadows, the bushes on either side were sparse and low so that we could have spotted a wolf or a bear well ahead. Thirty years later, the willows reached above my head for much of the way, making it impossible to see much more than the worn path at my feet.

CHAPTER 8

WATCHER AT THE DEN

"Wolves?" repeated the forestry ranger at Rock Lake in reply to my inquiry: "They poisoned eighteen of them here at Eagles' Nest Pass last winter. They didn't know there were that many."

Quite casually and matter of fact, he added that he had obtained this information from Gus the trapper, who had been hired by the Alberta Forestry Department to do local wolf control. Because of deep snow conditions, the man had not been able to get back to his bait station in the high country near the headwaters of Rock Creek until the end of March.

"When he finally got there, he was shocked what he found. Beside wolves, there were the carcasses of bears, eagles, and other animals. Dead ravens all over the place!"

This depressing story, dashing our hopes of finding wolves, greeted Irma and me at the start of our first hike into Jasper Park's backcountry. There was more to come. When we arrived at the Willow Creek district station, warden Norman told us that he had seen a pack of nine on Rock Lake last winter. "A couple of days later, Gus telephoned me to say that he had picked up seven dead on the lake. Poisoned!"

Fortunately, the warden also had some positive news. In late March, he had seen two wolves right in his yard. They were an odd couple. "One was small and grey, the other a great big black, at least a third higher at the shoulder than the other one."

Warden Norman was an old–timer, not used to showing positive feelings about wolves. In those days, the acceptable thing to say was that the only good wolf was a dead one. Yet, he could not hide his pleasure at having watched the pair right from his window. As his associates told me later, Norman had actually shot quite a few wolves in the park, which had been standard practice ever since Jasper's establishment. During the Alberta anti–rabies campaign of the 1950s, wolves had even been poisoned inside the park. The placing of strychnine–laced baits remained routine on provincial lands close to Jasper's boundaries until 1966, when a more tolerant attitude toward predators had begun to emerge.

As a wolf defender from day one, I like to think that I played a role in the final suspension of wolf control by government agents just north of Jasper Park. After the 1965 Willow Creek trip, I wrote an article for the *Edmonton Journal* and mailed a protest letter to the Alberta wildlife director. The poisoning was all the more regrettable, I submitted, since the predators were needed as nature's own check on the numbers of hoofed mammals in Jasper. The current overpopulation of elk had been denounced by several authorities. Already in 1947, destruction of winter range and poplar groves by elk was noted by biologist Ian MacTaggart–Cowan, who conducted the first wolf study by a Canadian government scientist. At that time, in an effort to dampen the reproductive potential of the elk herds, the wardens were culling several hundred cows each winter. Nevertheless, Cowan had recommended that the shooting of wolves be continued inside the park!

This contradictory attitude illustrates how difficult it was to buck the long–established trend. Biologist Cowan, who had published an insightful and objective report, might have hesitated to criticize park policy. Ironically, a hard–nosed wolf–killer like Norman had enjoyed seeing the predators. Or so we thought. Perhaps, we were mistaken and he had just been loath to show his true colours when confronted with a couple of newfangled wolf enthusiasts like us.

Unlike today's park wardens who live in Jasper town and only make occasional patrols into the backcountry, Norman still resided all year in his district. At the time of our visit, during the school holidays, his wife and children were there also. But over the long winter he had been alone, going back to town at certain intervals. Returning one day, snowshoeing along frozen Rock Lake, he had spotted the pack of nine wolves. The subsequent sighting of the pair by his cabin was another highlight, breaking the monotonous routine of his isolated existence and giving him something to talk about during his daily report to headquarters. Like most backcountry stations in those days, Willow Creek was hooked up to a telephone line strung from tree to tree. Wind storms could spell trouble. If the line went dead, the warden had to locate and repair the damage. Today the service makes use of two–way radios and satellite–phones.

When we told Norman that we had come here just to see and listen to wolves, he said: "We hear them all the time. Just west of here."

This was music to our ears and rekindled our high hopes. Although we neither saw nor heard any wolves during our three–day trip, we did indeed come across their fresh tracks on the horse trail west of the warden station. If I had known then what I learned several years later, we would most probably have found the local den occupied. For, when I finally happened to locate it, the weathered bones and skulls of deer and elk scattered about the site indicated that it had been used for a very long time.

~

The discovery of the Willow Creek wolf den came about by sheer luck in 1970, during a hot summer's day when I was hiking with Ludwig Carbyn, a biologist with the Canadian Wildlife Service, who had begun a three year wolf study in the area. In those days, before the routine application of radio–telemetry, Carbyn and his crew had to rely on footwork to keep track of the local pack. One day, when he had lost contact with the wolves, he and I investigated a large meadow some six miles (10 km) northwest of the Willow Creek station. In a muddy spot, we found the tracks of pups and we caught a brief glimpse of an adult wolf sneaking off through the bushes. Agreeing not to disturb the family, we retreated at once. On the way back to camp, while I was hiking through the woods to dodge the hot sun in the adjacent meadows, I stumbled onto the main den, which had quite recently been abandoned. It had been dug in a sandy bluff on a grassy opening littered with wolf scats and prey remains. There were three entrances to the den, the largest wide enough to allow a human to crawl in.

There was an alternate den a few hundred yards away, which was pointed out to us by Keith, the new district warden. His predecessor had shown Carbyn another set of dens a mile or two farther west. Evidently, generations of wolves had found the park–like landscape in the Willow Creek district to their liking.

After Carbyn terminated his study, which was actually a university project of limited duration, I had the country to myself. Each year, in early June, I checked out the dens for signs of recent use. One late afternoon, approaching silently through the woods, I halted as soon as the grassy bluff came into view. Screened by branches, something dark stirred on the mound of excavated earth in front of the burrow. Through the binoculars, I saw that it was a pup, a cute little bundle of black fur! Engrossed in its intimate world, it was swatting and snapping at mosquitoes that insisted on landing on its stubby nose. A minute later, the pup gave up the bother and went back underground.

Thrilled by the find, I retreated at once. Circling back through the trees I reached a low ridge which gave an open view on the meadow. The den was hidden just inside the woods opposite. I sat down on a shady log and prepared to wait. There was no wind, the whine of bugs the only sound. The low sun radiated heat like a furnace. Agonizingly slow, the fire ball sank below the rim of distant mountains. Soon after its blinding glare was extinguished, the meadow became a place of mystery. As dusk descended and the temperature dropped, mist began to veil the hollows. Through the glasses, I frequently scanned the forest edges expecting to see the parent wolves enter the stage at any moment.

Suddenly, the quiet was shattered by a series of explosive barks, much like an angry shepherd dog. It came from behind me, not far off, echoing harshly through the dark woods. Getting up, ill at ease, I glanced about but saw nothing but the trunks of trees. The barking was repeated and spurred me on to leave the way I had come, trying not to hurry until I was out of the woods. Arriving back in camp, the excitement of the encounter took a while to subside. I went to bed early, dreaming about the adventures tomorrow would bring.

Getting up at dawn, I made a fire and had an easy breakfast of oatmeal and tea. I then packed a few snacks, filled a thermos with hot water for instant coffee, and went on my way. Following the same route as the day before, I reached the lookout point on the wooded ridge and sat down, this time with the rising sun in my back. The night had been very cool, close to freezing in fact, and the mosquitoes had not yet awakened.

An hour or more passed but nothing stirred near the place where I knew the den to be, nor did I spot anything at all on the meadow. The nagging thought that the wolves might have left, perhaps triggered by my intrusion of the previous evening, kept revolving in my mind. The wolf that had barked so vehemently had clearly objected to my presence. If it had been a dominant adult, he or she might have decided to take the pups elsewhere overnight. As the morning progressed and the shadows became shorter, my worry increased. The urge to send out a call for information grew. I felt like a lonely wolf and I could not help acting accordingly.

Placing my cupped hands around my mouth, I was very conscious of the fact that what I was about to do would shred the sacred silence of this idyllic locale. Keeping my voice down, I let out just a short wail, a far cry from the real thing.

There was an almost instantaneous reaction. On the opposite edge of the meadow, several puppies bolted out of the trees and down the bank, bumping into each other and coming to a sudden stop. There were five of them, all black as coal! Looking in ecstasy into my direction, they raised their blunt snouts and opened up in an outburst of high–pitched yips, yowls, and squeals. Some even made a respectable try at keeping a sustained whine going. Just as abruptly, they shut up and sat down on their haunches, listening and looking about expectantly.

Beside myself with delight, I hardly dared to raise the glasses. Two of the pups sported a white star on their chest. Ambling about for awhile, they soon followed each other back up the bank and out of view.

The amazing thing was that there had been no response at all from any adult wolves. Evidently, the puppies were at home alone. Over time, I found this to be not uncommon, contrary to the popular notion that a so–called baby–sitter stays behind if the rest of the pack departs on a hunting foray. On several occasions I was sure that the Willow Creek puppies had been without supervision for many hours, perhaps even as long as a day and night. Of course, while all I saw were the puppies, any number of adults might have been lying up not far off, hidden from view. One year, there was a crippled adult present, possibly a female. Slender and skinny, she hobbled out of sight as I flushed her at close range. She had a bad limp in her hind leg and her spine was arched as if she had suffered a back injury, perhaps caused by a bullet or a blow from a hoof. Later that summer, after the pack had taken their pups elsewhere and I howled in the hope of getting a distant response, she let out the most pitifully sad cry I have ever heard. Abandoned, her fate seemed sealed. However, the fact that she had survived here for more than a month, proved that she had benefitted from the food brought to the den by the other adults.

~

On that first morning watch, the puppies did not emerge for a second time, and toward midday I left my post and returned to camp with the intention of trying again in the evening. When I got there, the sun had just slid below the jagged skyline and the meadow was in shade. To my surprise, the puppies were out in the open, scampering between the bushes or tottering down the bank of a narrow stream that transected the meadow. Their boldness was in great contrast to their timidity of this morning. And soon I discovered why: mom was at home! She was a silvery–grey wolf, asleep by the base of a large spruce. Perhaps she had just returned from a hunt and called the youngsters down from the den. Confident in her loving care, the puppies did not seem to have a worry in the world and were merrily exploring the surroundings. From time to time, the she–wolf raised her head and looked about, until she suddenly got up. At that same moment, the pups started running across the meadow, all into the same direction. The subject of everyone's attention turned out to be a large black wolf that had emerged from the woods. When he was met by the puppies, he touched their noses ever so gently. He then went back the way he had come, to be followed by the five little bundles of joy, stubby tails wagging. All too soon, the charming procession was hidden from sight. In the meantime, the silvery female had joined the group and she too disappeared behind a belt of trees leading to an adjacent meadow.

Ignoring the buzzing mosquitoes, I leaned back contentedly against the tree, savouring the wonderful memory of that happy family on their home ground. But my plans for another morning of watching were dashed. Overnight, a cold front moved in with steady rain. Packing up early, I hiked out, eager to cross tempestuous Rock Creek before its rain–swollen waters became too deep, and before the road out became impassable. Past experience had taught me to be cautious. Once, during a trip with my brother Fred who was visiting from Holland, the valley was swamped with torrential, all–night rains. The creek almost swept us off our feet as we forded the muddy torrent, but our hurry to reach the car proved futile. All four bridges along the Rock Lake access road had been washed out, stranding us for several days until repair crews arrived.

~

During subsequent trips, my watches at the den produced variable rewards. Sightings of the adults were usually brief. Most of their activities took place out of view in the woods or on the next meadow which was larger than the first but lacked a strategic hillside observation point. Instead of running after them, I kept my disturbances to a minimum by staying in the same spot. The wolves must have known exactly where I was. Perhaps, they had accepted my unobtrusive presence as harmless. If I went outside of my normal routine, it usually ended in a confrontation of sorts, with one or more wolves barking at me. Most of all, I hated to put them up unexpectedly from their hidden beds and spook them. They could be amazingly unguarded. One day, walking silently, all my senses on high alert, I inadvertently got so close to an adult and the puppies that I hardly dared to breathe. They were lying up in the grass. Afraid of causing a panic, I spied on them for no more than a few minutes and turned back as quietly as I had come.

By the same token, I will never know how often they sneaked up on me unnoticed as I sat against a tree. On several occasions, I was startled by a wolf making a u–turn nearby. It probably had come over to investigate a small noise I had made. On moments like that I felt sorry for the disturbance I had caused. I liked it best if the pack stayed well out on the meadow, unaware of me, while I watched them through the glasses.

The number of adults in the pack varied from six to nine. Interestingly, they differed not only in size and the colour of their pelage but also in build. Some individuals were long–legged and slender, others squarish. One brute stood out because of his huge head. Grey wolves could be distinguished from each other by the tail which might be dark–tipped. Black wolves were seldom entirely black. One year, there were two silvers with black legs, heads, and tails. The following year, two animals combined black extremities with brown bodies. These were the most beautiful wolves I have ever seen. In low sunlight, their fur glowed like burnished copper. The oddest wolf, as to the colour of its pelage, was the 1981 alpha male. He was exceptionally big and pure white except for a prominent, dark grey patch on his rump that looked like a saddle.

Because of their small stature, the pups were hardest to observe and easily hidden in the brushy vegetation that covered most of the meadow. By late July, when they were over two months old, they could spend the day anywhere in the long grass, between the bushes, or in the woods. Surprisingly bold and fearless, they explored widely. One of them showed up close to my tent, which I had placed in a secluded spot well away from the den site. Another day, as I was sitting on my usual lookout, a curious pup came up the slope to investigate. Just as it was to sniff my boot, the warden who was with me at the time attempted to snap its photograph and caused the pup to dash off in fright.

Most heartwarming to watch were the spontaneous interactions between the pups and the adults. New arrivals were welcomed with great enthusiasm by the pups. Ganging up, they reached up to lick the corners of its mouth, which stimulated the adult to regurgitate part of its stomach content. It happened so quick that I seldom heard or saw more than a retching cough and a glimpse of a reddish substance on the ground. With the pups swarming all over it, the food was gone in seconds. The tired animal would then flop down on its side for a long sleep, rebuffing frisky puppies with a snarl and a show of teeth.

At times, the wolves brought food in the raw state. The puppies spotted them much earlier than I. Once, while they were resting in the company of the silver female, the whole troupe took off all at once, as if the starting gun had sounded for the hundred yards dash. They met a big black wolf that came trotting up at the far end of the meadow. Looking proud, he was carrying a beaver in his mouth. Tails wagging wildly, the group surrounded and followed their benefactor into the woods. Later, I found the bucktoothed skull of the beaver, too hard to crush even for a wolf. The adults also brought the legs of deer and elk, which were treasured by the puppies, to be fought over or hidden away in a secret corner.

~

All too soon, the day arrived when the family left the natal den, and I watched and listened in vain. The once so vibrant meadow, where I had enjoyed such intimate sights, now looked lifeless and forlorn. Sadly, I wandered about for awhile, following the wolf trails that radiated out from the site, sign–posted with scats, but I seldom managed to relocate the wolves after their move. The district was just too vast and I generally stayed on familiar ground. The one place I always checked was the rendezvous site where Cedric and I had seen the pack a few years before. As the raven flies, it was less than two miles (3 km) from the den and both locations were linked by a network of game trails.

One September day, as a companion and I arrived at the Willow Creek station, the warden reported that just yesterday he had ridden his horse across the rendezvous site and scattered a pack of eleven adults and pups. Upon reaching the meadow, we sat down to scan the place, our hopes high. But except for the beauty of the vista, crowned by the jagged peaks of the Starlight Range, there was nothing to be seen. After a while we decided to withdraw and set up camp down the hill by a little stream, some distance away.

Returning by evening, we silently traversed the belt of trees leading to the edge of the meadow. Scanning the uneven surface through the binoculars, our hearts skipped a beat. There! Half–hidden in the grass, we spotted four sleeping pups, each one as black as coal. They raised their chubby heads and pricked up their ears often, but they remained at rest, unaware of our presence. Careful not to disturb them, we sat down, content to wait and watch.

Near the middle of the meadow was a small pond, reflecting the clear sky. A little farther away was a salt lick, a pool of trampled mud which was a magnet for generations of hoofed mammals that came here to satisfy their craving for soil minerals needed for the growth of bones and antlers. While we were watching, a great elk bull emerged from the woods and headed for the lick, where he began to nuzzle and chew the mud, not far from the four resting wolf pups. All they did was look.

As the evening progressed, the meadow was slowly sinking into shadow. Finally, one after the other, the pups stood up, stretching lazily. Two of them ambled into the direction of the elk, which swung his antlers over his back and trotted off, followed for a short distance by the pups. Did this giant bull really run from the baby wolves, we wondered? Or had he seen more dangerous foes? As we scanned the edge of the woods, our glasses focused on a large grey wolf emerging from the trees. He was met happily by one of the pups. Side by side, they loped into the open, to be greeted by the other pups with excited yaps and whines. The sounds proved a signal to other members of the pack, which must have been resting in the woods, for half a dozen wolves came running from several directions, eager to join.

One animal began to howl in a deep voice. Another chimed in on a different note and the pups burst out in an high–pitched clamour. While the wolves drew together in a tight huddle, tails wagging, the chorus increased to a rousing crescendo. Its vitality was so contagious that my companion felt compelled to join in. He imitated a howl. But it was a mistake, like an insult hurled at the faithful during their holiest ritual. The pack fell silent at once.

One of the wolves, a large and lanky black male, broke away from the group and started across the meadow into our direction. Halting some distance away, he barked threateningly, like an angry dog. To our relief, he then retreated to the edge of the meadow, where he paused again to stare at us. His hoarse barking reverberated against the hills. Presently, he turned and trotted off to rejoin the pack. Most of them had scattered and were howling intermittently from the woods. Soon, all were gone and silent. In the deepening dusk, we hastened back to our camp. Rekindling the fire, we talked in excited whispers until the stars came out.

Early next morning, we returned to the edge of the meadow to watch from the cover of trees. Not far away, near the pond, stood a mule deer buck, tail twitching nervously. When he bounded off into the woods, we lowered the binoculars and discovered that we too were being watched. A large grey wolf was standing on a slight rise just beyond the spot where the deer had been!

In a frontal profile, the animal looked oddly corpulent. His belly bulged well out to either side of his narrow flanks. Returning from a successful hunt, he had no doubt expected to find his pack on the rendezvous site. Twenty pounds of meat is not beyond a wolf's

84 ❧

appetite at a single meal and it was obvious that this animal's stomach was filled to capacity. He would have been eagerly greeted by hungry youngsters!

Turning away and trotting across the meadow, the wolf halted from time to time to raise his muzzle and howl. His voice had an odd ventriloquial quality. The sound seemed to come from a different animal, one much farther away. Also surprising was the fact that this very large wolf howled in such a high voice. His was the kind of sweet, questioning call that humans might translate into the phrase: "Here am I… Where are you…?"

Pausing frequently to repeat his lonely lament, the wolf reached the far side of the meadow where he disappeared into the forest. We continued to hear him at intervals, the howls becoming progressively weaker.

Wolves have much better hearing than do humans and they can contact each other six or more miles (10 km) away, especially under favourable conditions like this calm morning. We liked to think that this grey wolf, gorged with goodies, would soon be reunited with his loving family. Evidently, the pack had abandoned this rendezvous and moved to a new site, probably triggered by our intrusion of last night.

We camped for a second night in the same place and when it became clear that the wolves had not returned, we inspected the meadow. Tracks were much in evidence, especially in the mud of the mineral lick. The area was littered with scats and the bones of elk and deer, some old and bleached, others freshly gnawed. The pack seemed to have used this hang–out for a long time. Yet, not all of their prey species had fled this dangerous place. We saw four female deer, although none was accompanied by a fawn. The elk bull returned in the evening. When we imitated a wolf howl, he raised his antlered head and challenged us with a hoarse bugle. The deer did not react at all.

~

How do harmless denizens of the woods such as deer actually manage to coexist with an enemy as deadly as the wolf? This is a fascinating subject. In the grand setting of the Willow Creek meadows, I had enjoyed watching the wolf family at rest and at play, gentle, innocent, and caring for each other. There is, however, a very different aspect to their character. Their happiness comes at a price, paid in blood by other innocent creatures. Those that are least capable of defending themselves, such as the fawns of deer and the newborn of elk, are the first to fall prey to the wolf's slashing fangs. No doubt, I would have had a very different story to tell had I seen the pack in action, devouring a bleating moose calf or ripping the guts out of its courageous mother. These blood–curdling tragedies had been hidden from my human sensitivities, shrouded under the velvet cloak of the summer night, hushed by the whisper of aspen and the soughing of spruce. To obtain a glimpse into this nether world of tooth and claw I had to wait for another day, another place, and another season.

CHAPTER 9

TRACKING THE PACK

"We just seen your wolves!" the driver of the dumptruck shouted from his open window as he rounded the corner by the cabin. Without bothering to stop, he did not give us a chance to ask for details such as where and how many. The guy obviously knew why we were here. Two days earlier, on our hike in, we had briefly chatted with him or someone else in his crew who were widening a treacherous section of the bush road into remote Rock Lake. Just when the truck passed by, we were sitting on the steps of the forestry cabin, in need of a rest and enjoying the warmth of the March sun reflected off the snow. The previous day, we had exhausted ourselves on the trail of the local pack of wolves. And this was the second time that we had just missed seeing them!

The first time had been the day before, but under quite different circumstances. After camping out in the woods under the stars, Ron and I were eating breakfast by a good fire when the wolves howled from the nearby Willow Creek meadows. Ecstatic, we listened to their reveille, clear as a clarion call in the dead–calm of morning. After the howling stopped, we hurried through the forest toward the meadow, cautiously approaching the edge, fervently hoping and half–expecting to see the wolves in the open. Regretfully, they had gone. All we found were their tracks.

One animal had actually entered the woods, proceeding into our direction and retreating again quite close by. Meanwhile, oblivious of being stalked, we had been eating and chatting. The scout had reassembled with the pack and they had sent us their emphatic verbal message: this was their territory, their domain. We were but intruders, strangers in paradise.

The disappointment of just having missed them hit us all the harder since we had intended to bivouac on the edge of the meadow, precisely in the hope of seeing wolves in the open. The previous night, tired out by the calf–deep snow, we had given up just short of our goal and made our camp in the shelter of the woods. Apart from the ever–cherished dream of watching wolves, our intention had simply been to locate tracks and follow them for two days, carrying our gear on our backs and sleeping wherever night fall would overtake us. But the first day we had drawn a blank in the freshly fallen snow. Now that the wolves had found us, we would be on the hottest trail imaginable!

~

Detouring around us and leaving the meadow, the pack had taken a direction toward Rock Lake. This was just great with us since it would shorten the way back. However, instead of staying in the valley bottom, the wolves had ascended the heavily wooded mountainside. Travelling in single file, stepping exactly into each other's prints so as to save energy, they threaded a path along the contour of the slope. We had no trouble following, slowly plodding along. We did not speak, each one of us thinking his own thoughts, enjoying this long–awaited thrill, trailing the masters of this wild valley. We were treading on virgin ground where no human had set foot before, at least not in modern times. Under the protection of the national park act, the forest on this mountainside had not been touched by people for nearly a century.

The wolves seemed to know where they were going, probably following an ancient game trail, gradually gaining in elevation. Now and then, through an opening in the trees, we obtained a perspective on the valley below. After a few miles of steady travel, the pack had come to a stop for a rest or some play as evidenced by a trampled area. Near its centre lay the front leg of a deer, freshly gnawed. How did it get here, we wondered? Perhaps a young wolf had carried it along from its last meal, until it had dropped its toy, like a tired child. We wondered how long the pack had dawdled here, allowing us to gain time. The idea of catching up and seeing our elusive friends just ahead never left our mind.

Pushing on, we again encountered a place where the pack had come to a stop, huddling together under a gnarled pine. It looked like they had sensed prey ahead, for when they left again they did so at a gallop. This could be interesting. The ultimate objective of our quest was to check out hunting activity, to find the fresh remains of a prey and unravel the wolves' strategy in making their kill.

Bounding along, the pack had spread out in an uphill direction. Unable to follow more than one spoor at the time, we occasionally backtracked and double–checked, trying to figure out what was going on. It was a tiring and time–consuming puzzle. Eventually, we found the fresh beds of two animals which had apparently taken off in great haste. At first we thought that they were deer, but instead they proved to be other wolves, trespassers on the pack's territory! The sleeping pair had been rudely awakened by the arrival of the pack, its intent far from friendly. Lucky for them, the interlopers had escaped, and the pack had resumed its original direction in single file.

A little farther on, we were confronted with a steep ravine. Going down might be easy for a loose–limbed wolf, with four legs to steady itself, but for us, top–heavy bipeds, the descent was hazardous. Clambering up with equal difficulty on the other side, we found that the pack had changed direction and turned uphill, following the draw. As the gradient became steeper and the snow thicker, we were forced to stop frequently for brief rests. In the end, sweating and quite exhausted, we sat down to reconsider our plans. Did it make sense to go on? By the looks of it, the wolves were heading for the summit of the Bosche Range which towered up to timberline, an elevation of some 3200 feet (960 m) higher than the lake. On the other side of that mountain range lay the secluded Moosehorn Valley. There, the pack could either choose to turn south, back into Jasper National Park, or go the other way and descend onto provincial lands. From our point of view, the journey would be too taxing, and the farther we proceeded today, the longer tomorrow's return.

As mountain wanderers, wolves clearly outclassed humans by a wide margin. There had been no contest. We were forced to give up and all we could do now, was to wish them *bon voyage*. Although we felt sorry to let them go, we were quite satisfied with the trip. Despite the short distance covered, no more than four or five miles (6–8 km) in total, it had been an intriguing and rewarding experience.

Taking the shortest route down, we plowed through deep snow until we reached Rock Lake where the going was easier. That evening, instead of roughing it another night in the bush, we made ourselves comfortable in the warm forestry cabin. Tentatively, our idea was to go back up the mountain tomorrow, without backpacks, and resume tracking where we had left off. However, we had risen late and still felt quite tired. After breakfast, while we were lazing in the sun on the cabin porch, the truck driver had come by and blurted out his cryptic message that echoed in our ears: "We seen your wolves!"

But where? The answer came to us an hour or so later. Not far from where our foot prints of the day before emerged from the woods, the wolves had come back down the mountain! They probably had done so this morning, just when the construction crew drove by on the road above the cabin, while we were sitting on the steps. Unfortunately, our view of the lake had been blocked by trees.

Although it was deeply disappointing to realize that we had just missed out on the sight of a pack of wolves gambolling on the lake, it was great to see their fresh tracks right overtop of our own. By the looks of it, they had spent quite some time on the snow–covered ice, playing and running about. Upon leaving the lake, they had entered the woods on the opposite shore and gone into a westerly direction. Lacking the energy and the time for further pursuit, we turned back soon and made a lunch fire on the shore, relaxing and savouring the view of the wintry mountains before heading back home to the city.

As to the exploits of the pack over the past twenty–four hours, we could only guess. Going by the amount of scat the wolves had left on the lake, it could well be that they had eaten well. Perhaps they had chased and killed a deer high on the ridge? Or had they gone all the way down into the Moosehorn Valley to hunt elk? We will never know.

One way to identify the kind of prey the wolves had eaten would have been by collecting their scat and examining the furry contents under a microscope in a laboratory. That sort of research, which lay beyond our intentions and ability, was at that time the task of two teams of government biologists, who were simultaneously studying this very same pack. On this particular weekend, we had happened to be somewhere between the respective study areas of these two teams and we had been in a position to contribute our small share of information. Firstly, we had determined that the pack had eleven members. In the woods, it had been next to impossible to figure out the exact size of this large pack, but today, after their descent onto the snow–covered lake, the animals had spread out and allowed us to count individual spoors. As we heard afterwards, our trip had coincided with field work by Ludwig Carbyn, who had tracked the same pack of eleven in Jasper Park's main valley the day before. Furthermore, just after we went home, the fresh trail was picked up west of Rock Lake by a research team of the Alberta Wildlife Division. Driving snowmobiles, they had followed the wolves for about twelve miles (20 km) along the bush trail into the Willmore Wilderness. At Eagle's Nest Pass, the pack had crossed over into the Rock Creek valley and turned south, heading back into Jasper National Park. This meant that the wolves, in a couple of days, had travelled a circuit of well over one hundred miles (160 km), an indication of the vastness of their territory.

Of course, today, similar information is much more easily gathered with the help of sophisticated telemetry. Wolves fitted with radio–collars hooked up to satellites send a steady stream of data to the biologist, who does not even have to leave the office to know where the animal is going. Once every day or so, its whereabouts are transmitted and displayed on a computer screen, complete with GIS information on habitat and other factors. However, the figures and symbols are barren of a host of intimate subtleties, such as we had enjoyed during our trip. No pulsing computer blip can ever capture the magic of how the pack had assembled under the gnarled pine overlooking the valley. Neither can it show us how a wolf bounds with giant leaps down a snow–clogged ravine or up a sun–lit hillside in pursuit of enemy or prey. No technical wizardry can ever replace or remotely approach the ambience and the beauty of the winter woods such as Ron and I had seen with our own eyes and felt in our tired bones on that March weekend long ago.

Chapter 10

Black Bear Trouble

"Tourist becomes bear's Dutch treat," ran the jocular headline in a June 1996 Alberta newspaper. It reported on a hilarious yet potentially serious confrontation in which a black bear got the better of a pushy photographer from The Netherlands. When the man saw the animal cross the road and vanish into the woods, he had parked the car, grabbed his camera, and went after the bear on foot. Apparently, it objected to having its picture taken for it made several bluff charges, stopping just short of physical contact. Eventually, losing his nerve, the man raced back, frantically darting between the trees. Just as he reached the car, his anxious wife opened the door, while the pursuing bear sank its teeth in the buttocks of her screaming husband.

The couple drove to the hospital in Jasper where the man received several stitches to close the puncture wounds. However, the park wardens did not plan to take action against the bear, for in their opinion it had done nothing wrong. "The fellow got too close, into the animal's personal space."

Be that as it may, the question of why some bears charge instead of flee at the approach of people, remains as puzzling as ever.

The above incident could have easily happened to me four decades ago when I was a new immigrant from the same country as the headline–making tourist. Consumed by a similarly impatient desire to see bears, I too was naive enough to make them the butt of my unwanted attentions. For instance, when a sow with three cubs crossed the Banff Park highway, I hurriedly parked the car on the shoulder and took off in pursuit, camera in hand. I did, however, keep a respectable distance and my objective was primarily to watch the family. The bear made a stand on the crest of a slope. Her cubs were out of sight, probably sent up a tree. Broadside, she was waiting for my next move. I shudder to think of what could have happened had I taken another step. Downhill, she would have been on top of me in an instant. Cautiously, I retreated.

Another time, when I shadowed a big randy boar on a stroll through the forest, my impudence might easily have been punished in a very painful way. Glancing over his shoulder at me, the bear rode down a sapling and rubbed his crotch. Eventually, he dropped out of sight behind the upturned roots of a blow–down, perhaps for a nap or to lie up in ambush. I wisely decided to leave him alone.

I hasten to add that I soon gave up such brazen tactics. At the time I had convinced myself that bears would not harm people who did not intend to harm them. I now view this guileless attitude, still prevalent in some quarters, as wishful self–delusion. Yet, animal trainers with an intimate knowledge of bears credit them not only with intelligence but also a finely tuned sixth sense that allows them to pick up on a person's feelings toward them, either loving or hostile. Of course, this should come as no surprise. Dogs demonstrate the same faculty, and in fact, so do people. Aggressive intentions are unconsciously betrayed by body language.

As to bears, I once saw a telling example of their ability to distinguish between friend and foe. During the early 1960s, when it was still common practice for tourists to feed animals along park roads, I stopped to join a guy who was taking photographs of a sow with tiny cubs. The man had obviously given her food, which she was eating on the road shoulder. As I walked up briskly, carrying camera and tripod, I happened to notice that she had quit chewing and flattened her ears. Backing off right there and then, I returned to the car and drove up closer to the other tourist. Getting out gingerly, I placed the tripod on the pavement and bent down to look into the viewfinder just as the sow's ears went flat again in preparation for the charge. I quickly took cover inside the car.

Why did she mistrust me while tolerating the other guy taking pictures at very close quarters? The difference was that he had given her food and I had no such intention. In fact, I was strongly against such practices. No doubt, these negative vibes had been sensed by the sow, ever ready to attack anyone perceived as a threat to her cute, playful cubs.

Feeding wildlife has become illegal in all national parks, and to counter the ingrained human temptation of handing out goodies to a life–size teddy, park managers promote the catchy slogan: "A fed bear is a dead bear." For, once the animal becomes dependent upon handouts from tourists, it turns into a spoiled brat that may have to be destroyed to protect the public who caused the problem in the first place. During the first year of enforcement of their new policy, park wardens not only handed out tickets to offending people, but they shot dozens of panhandling bruins in both Jasper and Banff.

Today, bears that highjack groceries in public campgrounds or scavenge garbage in the townsite are either trapped and released into the backcountry or "removed." The latter is a management euphemism for killing the bear. However drastic this final option may seem, particularly in a national park where wild animals are supposed to enjoy full protection, the more humane method of moving the offender elsewhere has proven to be half–baked. In one recent case, a Banff bear that had been trucked twenty–two miles (35 km) out of town was back the next day. A Jasper bear, netted and slung under a helicopter for a flight of thirty–two miles (50) km), took a few days longer but zeroed in unerringly on its old stomping ground. These are by no means record distances. In August of 1998, a bear transported from Banff to Kootenay National Park found its way back across 120 miles (200 km) of mountains. The end result was that it was "removed" in a more terminal way.

There are other problems with transplanting. Any vacancy thus created is usually quickly taken up by intruders that may cause similar trouble. For that reason, wardens now prefer a very different strategy called hazing. Bears that show up near roads and tourist facilities are shot with rubber bullets so as to increase their shyness. This makes sense to me. Throwing things at animals you do not like comes naturally to most people. Many years ago, when Irma and I were camped on a popular site in Jasper, another party was approached by a black bear that made off with a large raw steak. Without thinking, I immediately ran after the thief and let go with a barrage of sticks and stones. To my astonishment, son Richard, barely three years old, had followed daddy's example and stood by my side, also shouting and throwing things at the big bad bear. Lucky for us, the culprit did not retaliate!

Another negative consequence of transplanting troublesome bears has to do with their own chances for survival in unfamiliar country. There is very little evidence that they do well in new surroundings where they may run into an unkind reception from territorial residents of their own sort. Moreover, the objective of reducing human–bear conflict in the one location can backfire for people in the other. To a hiker, the loss of food to a spoiled campground raider, that has been dumped into the backcountry, represents a calamity, especially if the hiker still has several days to go before the groceries can be replenished.

This is exactly what happened to three young men I found sitting dejectedly in a campground west of Willow Creek, where I had intended to stay as well. They were halfway in a ten–day trip and the night before, one of the guys had lost his backpack. It had been torn down out of a tree by a black bear that showed no fear of people. Most likely, the culprit had been flown up here by the authorities from the main valley. At that time, the North Boundary district was a common destination for bothersome bruins. Unsuspecting backcountry users were left to fend for themselves. In those days, few if any campsites had bear–proof facilities for securing food at night. This was then not a high priority since truly wild bears generally avoid humans. By contrast, the transplanted bandits were habituated to people and had brought their bad habits with them. Upon release, they could easily locate the widely–scattered backcountry campgrounds by the smell of wood smoke, which they had learned to associate with food.

Irma and I saw a startling demonstration of this during our very first hike into Willow Creek. We had just crossed Rock Creek and were eating lunch in a sunny spot on a wide–open gravel bar. Our wet boots were drying by the fire. Suddenly, a small black bear emerged from the trees and came straight toward us, its nose into the smoke that drifted low downwind. Jumping to our feet, we screamed and waved our arms. To our great relief, the bear turned back the way it had come. Presently, it tried again and we had to repeat our vehement protest. Retreating under a hail of rocks, it stayed back but sneaked about for awhile in the adjacent bushes, making us feel very uneasy indeed. What if we had panicked and fled, leaving our packs on the ground? At the time, I found it difficult to believe that a wild animal would behave in such a fearless way in a location far away from human habitation. It was not until much later that I recognized the possible connection between our predicament and the so–called humane bear removal practices of our friends the wardens, who might have given this cheeky raider a free helicopter ride into the backwoods.

Since that time, I have had several other run–ins with black bears around my camp. One day an astonishingly bold intruder looked at me from behind the tent as I was pulling on my hiking boots. This smallish yearling was easily shoed off, but nevertheless I decided to leave at once. A similar surprise befell my buddy Brian, who woke up from a midday snooze just in time to see a small bear carry off his backpack, which had been leaning against a nearby tree. When Brian shouted, the thief dropped the pack. I suspect that both incidents involved transplanted rogues, although there is another possibility. It might just be that these bears were completely ignorant of humans for the simple reason that they had never come across one before. Or perhaps they were the offspring of a transplanted sow who had failed to teach her youngsters that people were to be avoided. As with other wild animals, there are always more questions than answers. Bears will continue to surprise us. The word most often used to describe their character is unpredictable.

In retrospect, I consider myself very lucky to have suffered no serious trouble with fearless bruins over three decades of backcountry hiking. I seldom even bothered to hang my pack high in a tree. Instead I just placed it against the trunk a couple of steps from the tent. The only critters to touch my precious supplies were squirrels or porcupines. Not a single bear has ever run off with my food. Of course, the risk is always there, and before getting up in the morning my first look outside always concerned the backpack. To see it still standing in the same spot where I had left it the evening before, made for a reassuring start of the new day.

My night–time worries were most acute if I had placed my tent in the woods out of sight of the trails. This is what I had been advised to do to get around the official camping restrictions. I had obtained this privilege from the district warden so as to be closer to the wolf dens, which were at least two miles (3 km) away from a public campground. However, if I had ended up in serious trouble no one would have known where to find me. Fresh bear scats were often in evidence in the vicinity, for scavenging bears are attracted to wolf home–sites by the smell of food items that wolves bring to their pups. One morning the pack chased a big boar along the edge of the woods. Another day, I found the hind leg of a pup, perhaps killed by a bear. If I heard a wolf bark at night I assumed its protest was directed at an unwelcome, four–legged visitor. Having a mutual enemy felt like a common bond between the wolves and me.

~

As to the average frequency of bear problems, compared to the past, a hiker's chances of escaping confrontations may have declined in reverse proportion to the rising number of people using backcountry campsites today. The two worst experiences were my last. One of them took place at Celestine Lake, a popular hike–in campsite at the lower end of the North Boundary Trail. One fall evening, when Arne and I shared the grounds with four or five other parties, a shout rang out: "Bear!" Looking up, we saw a large boar calmly walk up to a couple of campers who were having supper. The guys fled and the bear confiscated their food. He then climbed a tree in which someone else had hung a backpack. Tearing it down, the boar scattered the contents, consumed all that was edible, then ambled off, leaving everyone in an apprehensive state of mind.

As it was too late in the day to go elsewhere, Arne and I decided to make the best of it. At least we were not alone here. After all, humans are social creatures that derive comfort and strength from each other's company. Before turning in for the night, we took the necessary precautions such as securing our food in a plastic bag and hoisting it up with a rope thrown over a high branch. Freely suspended in the air, no bear would be able to touch that. I placed my backpack, empty this time, against the bowl of a spruce, and Arne hung his from a branch higher up. It too was supposed to contain no food at all. However, he had forgotten to check one of the pockets which contained, as we found out later, a plastic jar of sugar!

Darkness was falling and we were enjoying a fireside chat with some other guys when one of them pointed and said: "Hey, look! He's back!"

To our dismay, the boar was right by our tents, smelling and tearing my pack. Finding nothing there, he stood up on his hind legs and seized Arne's pack. He then lay down and began licking the sugar that had spilled from the container onto the ground. Indignant and reckless, I ran up, shouting loudly, but I checked myself just a few steps short of bumping into the big brute. Fortunately, he ignored me. He could so easily have flattened me with one swipe of his mighty paw! On moments like that, when one feels afraid as well as angry, it is easy to identify with the hunters among us who would have known what to do....

Quite helpless, we waited for the boorish visitor to finish and leave. By the time we were ready for sleep, a full moon had risen in the clear night sky, and the ghoulish calls of a barred owl resounded from the shadows of the forest. As is my habit, I left the entrance of the pup tent partly open with the bug screen zipped shut, and as usual I slept fitfully. Around midnight, when I awoke, the moonlit campsite was almost as bright as day and the first thing I saw was the huge dark bulk of the bear, a few feet in front of the tent! He was again indulging his sweet tooth and nuzzling the ground. I dared not move a muscle until he moved off.

By the time I awoke once more, the moon had gone. It was pitch black outside and the bear was back again! Petrified, cold sweat breaking out all over, I listened to his sniffing and smacking just a few feet away from my prone head, only the bug screen in between. Right there and then, I vowed never to use popular campsites again!

Not long thereafter, I had an equally shocking and perhaps even more frightening experience at Willow Creek. The site is still as remote as ever, about nine miles (13 km) from the trailhead at Rock Lake, but during the summer months it attracts a growing trickle of visitors. To have the place to myself, I had gone there in fall. I had met no one on the trail and the warden station had been vacant. As a highlight of the brilliant afternoon, I had wet a line in a beaverpond along the way, and a fine rainbow trout had risen to the fly. After frying the fish to a delicious treat, I had taken the usual precaution of throwing the bones onto the coals. And the tent was a safe distance away from the fire box.

Upon retiring, I set the backpack against the base of a spruce a few steps away. The wind was dead calm. Sometime near midnight, I was startled out of my slumber by a muffled blow, followed by huffing noises coming from the direction of the fire box. Bears! Judging by sound, there must have been several, probably a sow and her cubs, bickering together. Presently, in the silence that followed, I became aware of furtive footsteps by the slight creaking of tinder dry needles. A bear was padding right up to my tent! I panicked, screamed and cursed, banging the canvas: "Get away, you damned bears! Get lost! *Zodemieterop!*" The latter is a Dutch profanity that I use only under the most severe provocation. Luckily it worked…!

Listening to the calm of night, fearful of every rustle and snap, I stayed awake for a long time. Turning over in my bag, I drifted back to sleep, and when I awoke again, it was light. Oh, the intense joy of being alive! Of standing up on healthy limbs, my only means of getting out of here, out of this lonely, dangerous place….

The backpack was still there, leaning against the tree, untouched. In the invigoratingly cool of morning, I walked up to the fire box. The ashes had been scraped out. Obviously, the bears had sniffed out the charred pieces of trout backbone. The wood block I had used for a table yesterday had been knocked over, which explained the muffled bump that had woken me up in the night. Other than that, there was no other sign of the bears, no tracks, no scats. The experience, although it all ended well, left me badly shaken, and I have not used my pup tent since.

~

Meeting bears along the trail is quite a different matter. At least you can see them. You just never know when! Silent and sneaky, they may come and go unnoticed. Walking along in the woods just as quietly, I once saw the head and shoulders of a big boar rise up from behind nearby bushes. "Hi there!" I said in a calm voice, standing quite still. For a few tense moments, he looked at me, then dropped back onto all fours, vanishing as abruptly and silently as he had popped up.

The most rewarding sightings are those in which you spot the bear before it becomes aware of you. You then have the choice of either leaving unobtrusively or of stopping to watch and learn something about their habits. Eating is what bears do for a living, so to speak, and they are true gourmets. One moment they may delicately sample a flower or berry, next they use brute force to rip open a log or turn over a boulder to get at the insects underneath. Not even an ant is too small to be licked up and savoured. If they detect your scent, they wave their nose in the air, and hurry off. As a rule, they are very shy. But not always.

On a twenty–five mile (40 km) bicycle trip up the Snake Indian Valley, my teen–aged son Richard and I met a gigantic black boar that refused to flee. He stood his ground and no amount of shouting and arm waving seemed to impress him. The best way to deal with this situation might have been to give him the right of way, but the trail traversed a steep slope and we did not feel like entering the dense woods on either side. Nor did we want to retreat. After some fifteen minutes, the bear diffused the situation by taking a few steps into our direction, nibbling the vegetation, then turning off into the bush as if he had intended to do that all along. Elated to see him go, we did not feel secure until well past the spot where he had vanished.

Continuing down the trail, Richard was the first to round the next bend, and again I heard him shout: "Dad, a bear!"

And indeed, some distance ahead, another huge boar stood looking at us. This one also waited many minutes before deciding upon a course of action. To our relief, he ended the standoff in the same circumspect manner as the first one, by ambling a few steps toward us, then turning off into the woods. This was their way of saving face. The double scare happened to us on a hot June afternoon when we had not expected bears to be active. These two males were likely abroad for a good reason. It was mating time and they were on an urgent and visceral mission. Perhaps both were following the alluring scent of the same sow that had passed this way earlier.

~

Although she is smaller than the average boar, the female black bear may be the most temperamental. With cubs to protect, she prefers to stay out of the way but is easily aroused to aggressive fury if she feels cornered. Her razor–edge mood is an innate defence against her own kind. Amorous boars can be a threat to her cubs. They might even kill and eat the little ones. Their loss would bring the sow back into heat, making her available to the randy suitor waiting to sire her next brood.

Sows with small cubs tend to withdraw into areas that provide good cover, usually well away from hiking trails. But you never know. One wet morning, Brian and I were following a narrow path between tall willows. Upon emerging into a small opening, we saw a black bear watching us from the opposite edge, a mere thirty yards ahead. Next moment, as she turned away into the dense woods, we caught a glimpse of her cubs. A chance encounter such as this might well have ended quite differently had I been alone and silent, taking the bear by surprise. Fortunately, we had been talking together, which had alerted the sow, preparing her for flight.

Another day I narrowly escaped a close encounter of the wrong kind by the skin of my teeth, by sheer happenstance, or at the divine direction of my guardian angel. The memory of it makes my skin crawl. I had been snoozing in a shady spot, waiting for the cool of evening, when, for some reason, I got to my feet. Instantly, two things happened at once. A little distance off in the woods, a bear snorted explosively. Much closer by, four tiny cubs scampered up a tree with the speed of a rocket. The sound of their claws on bark was as loud as a drum roll. Fortunately the sow stayed where she was, and I backed off to a respectable distance, until she called her beloved babies down and ran away.

Apparently, mother had been following behind the cubs, which had advanced quite close to me. What if I had not chanced to stand up when I did? Half a minute later, the unsuspecting little ones would have been on top of me or even past my prone figure. Had I stumbled to my feet then, the sow would have knocked me out for sure.

~

As if we did not already have enough to worry about, what with dominant boars and defensive sows, there is the so–called predatory black bear. This is the rarest category of *Ursus americanus* but the most dangerous. It actually hunts down humans as prey! The term "predatory black bear" was advanced by Canadian expert Stephen Herrero, who has documented bear attacks on humans over several decades. However, all bruins are opportunistic predators that jump at the chance of dining on defenceless creatures such as deer fawns and moose calves. Bears even manage to catch adult hoofed mammals. The reason that some individual bears treat humans as fair game may have to do with ignorance or extreme hunger. Fact is, unprovoked attacks on people occur most often in wilderness situations. The perpetrators are probably bears that are not familiar with the human species. Other serious incidents have involved old and starving individuals. To them, clumsy bipeds may look like easy prey. Unless armed, we stand a very slim chance indeed of beating off a determined predator.

During the past century, according to official figures, black bears are known to have killed some thirty people across North America. This can only be an absolute minimum. Who knows how many others, reported missing in the backwoods, have actually fallen victim to the murderous, man–hunting beast? Solitary anglers, quietly working a creek and concentrating on the water, have proven to be vulnerable. If the victim is dragged into the bushes, no–one may ever find the evidence. Several recent cases took place in Alberta. Just east of Jasper park, the partly consumed corpse of a missing fisherman was recovered after an intensive search. A nearby black bear was shot and its stomach contents included traces of the man's clothing. In another gruesome case, an outdoorsman from Beaverlodge, Alberta, was pulled out of a tree and killed. The man's companion had fled to get armed help. When they arrived, the bear was burying the mangled body and rushed out of cover to charge the other men. The aggressive beast was felled with a blast from a shotgun at point blank range.

During a five year period, 1994–1998, I collected over two dozen newspaper clippings reporting on marauding black bears in Canada. They include at least six human mortalities, each one a gruesome tragedy. A couple, camping on an island in an Ontario lake, was dragged out of their tent and partly devoured. Apparently, the man had put up a futile defence with an axe. In British Columbia, an unarmed rancher was stalked and killed on his cow pasture. Searchers shot the bear which was still with the body. In that same province, a black bear entered the backyard of a rural house and mauled a four–year–old boy while his mother put up a heroic fight using a shovel and a flowerpot. The intervention of neighbours eventually saved her life but not her son who died in hospital.

In August of 1997, at the Liard Hotsprings along the Alaska Highway, a female tourist was seized and killed by a skinny black bear that suddenly emerged from the bushes. Another tourist, a courageous gentleman, vainly came to the woman's rescue and was also killed. The bear subsequently turned on a young man, knocked him down and began eating him alive, tearing the meat of the victim's legs! The monster was eventually shot by an American tourist, who had run to his vehicle and returned with his rifle.

In addition to these lethal encounters, Canadian papers carried news stories of at least eight other serious and potentially fatal attacks in Yukon, Ontario, Manitoba, British Columbia, and Alberta. They involved campers, forestry workers, biology students, and hunters. In one graphic case, the attendant of an isolated fire lookout station in northern Alberta happened to see a bear rush out of the trees and seize his girlfriend, who had been doing chores outside. Kicking and screaming, she was dragged into the bushes. Fortunately the man had a gun and saved the woman's life in the nick of time. The bear proved to be a very skinny and mangy individual.

While I was writing this chapter, the local paper ran news items about two more unprovoked attacks. In Jasper, an eleven–year–old boy who was fishing near the townsite, was pursued and bitten by a young bear. The kid received a scratch on his head but managed to escape on his bicycle. On a campground in southern Alberta, a mother and her young child were bitten and dragged out of their tent in the middle of the night. Their screams alerted another camper who beat off the bear with a stick.

In addition, there were news items about so–called nuisance bears that invaded backyards and terrorized people in rural and even urban settings of western Canada. For instance during the summer of 1996, thirty black bears were trapped and trucked out of the outskirts of Vancouver. In 1998, when natural foods such as berries were scarce, well over a thousand invading bears were shot in towns all across the interior of the province. Specific figures for the town of Fort St. John and the Okanagan valley were respectively forty and forty–seven. Near Prince George, twenty–eight bears were shot on one oatfield! In some years when berry crops failed, hungry bear warnings have also been issued for the Jasper and Banff National Parks.

The topic always makes for fascinating news, but it certainly does not contribute to the peace of mind of the mountain hiker. What can we do if our worst nightmare comes to pass? As far as the black bear is concerned, climbing a tree would be useless, even dangerous, since the animal's sharply curved claws are designed for arboreal exploits. Black bears like to eat poplar buds in spring. They also climb trees to escape from an even bigger bully, the grizzly bear. The latter's long, flat claws, used for digging into the ground, are not suitable for climbing.

To save oneself from the predatory black bear, a person's only hope, according to Stephen Herrero, is to fight back. People under attack have been known to stick their fingers into the bear's eyes or to tweak its nose. The final outcome of the assault may depend on any number of variables.

While prevention is the key word, how can we best protect ourselves? Even guns, which are not permitted in the National Parks anyway, offer no guaranteed way out. Many a lone hunter has been attacked before he or she had a chance to aim. My good friend Brian used to carry a flare, designed as a last resort for people in distress. It is a pen–sized cylinder that shoots a fire cracker high into the air where it explodes with a loud bang and a bright flash. Problem is, if aimed at a charging bear, the projectile might detonate well behind the animal. This could have the opposite effect of scaring the bear away. It might instead flee toward the person who fired the shot.

The most practical deterrent that really works, if engaged at the right moment, is the pepper spray can. It can be strapped onto one's backpack or belt. But like a gun, it too requires a few hurried seconds for proper employment. One needs to get it in hand, remove the safety, aim in the right direction and press the trigger! The pressurized contents shoot a jet of liquid. Under calm conditions, it carries about six to ten paces. However, when I tested mine during a strong cross wind, I could not even hit a stump three steps away. Against the wind, one might easily knock oneself out. Under duress, people have indeed been known to accidentally spray themselves, a fate almost worse than death, if you ask me.

Thusfar, I have taken the can into my sweaty, shaking hands half a dozen times, but the need to actually press the trigger never arose. One day, as I was standing on one side of the railway bridge across the Snake Indian River, a bear started across from the other end, about a hundred yards away. Spray can at the ready, I waited for the moment of truth. Here and now I would prove to myself that I had the cool–blooded guts to deal with this dreaded of all emergencies. The test never materialized. For some reason, before he was halfway, the bear turned back and disappeared on the far side of the bridge.

Relieved and feeling invincible, as if I had achieved a major victory, I strapped the can holster back onto my belt and crossed the bridge on the narrow catwalk alongside the rails. The sound of my steps reverberated through the metal structure. Just as I emerged on the other end, I saw the bear standing around the corner of the gangway, only the twinned tracks between us. Next moment, it withdrew into the bushes while I was still fumbling for my can.

If this bear had meant to harm me, I would have been too late to stop the charge. The question remains why the animal had not fled upon hearing my loud, approaching footsteps. The probable answer is that it was waiting in ambush, expecting a deer or elk. Black bears are opportunistic predators that kill any large mammal they can overcome.

The only time that I actually used the pepper spray was quite by accident and the unwitting target was myself. This backfiring episode took place when I had obtained permission to stay for a few days in a backcountry cabin which had been secured against bears by an electric fence powered by a solar panel. The reason why this extra protection had been installed was that the place had recently been broken into. A black bear had ripped out part of the plywood wall and trashed the interior. After the cabin had been repaired, a technician had surrounded it with the fence that featured a simple gate for the convenience of the few government biologists using the remote facility. By unhooking a wooden handle, one could detach the live wire from a post and step through the opening. Upon leaving in the morning, I did not bother to open the gate and just stepped over the fence. I paid dearly for my laziness....

The night had been cold. The liquid propellant in the pepper spray is not supposed to freeze, so I was carrying the can in my trouser pocket. By coincidence, I had recently lost the plastic safety catch and replaced it with a chip of wood. Just as I lifted my long legs over the fence, the improvised safety became dislodged and the trigger was accidentally pressed. The can fired in my pocket with the hiss of an angry snake! It took a few seconds before I realized what had happened. Then, an excruciating pain flared up in my crotch. Hurrying back inside, I took my pants off, made an effort to wash them, and changed underwear. In the process, my hands got thoroughly contaminated and whenever I touched my eyes or nose they began to burn. It spoiled the remainder of my stay.

On a subsequent trip, the same thing happened again when all I did was sit down on a log. The bears must have been laughing!

CHAPTER 11

GRIZZLY ENCOUNTERS

"Hey, wait a second!" my buddy Paul said in a soft but urgent voice. Hiking behind me on the narrow trail to Willow Creek, he had stopped. His face was aghast.

"What?"

"A bear just turned around right in front of you! I saw his fat behind."

Striding along at a brisk pace, I had been looking for animal tracks in the newly–fallen snow at my feet. Incredulous, I walked a few steps farther along the winding woodland path until I came across the very fresh pads of a bear, coming and going! Quite excited now, I searched for detailed imprints in the thin layer of white fluff. "Look!" Pointing down at the fuzzy but distinct evidence of long claws, I added: "It's a grizzly all right."

Surprised by my matter of fact manner, Paul, still very subdued, wondered: "You're not afraid?"

"Why should I be? This guy is obviously more afraid of us than we are of him. He turned back, he gave us the right of way."

Of course, it was easy for me to talk. I had been spared the frightening shock of seeing the beast so close up, a prospect I had dreaded each time I had walked this section of trail that passed through a rich stand of buffalo berry bushes. Their shiny, bright red fruits are a favourite food for bears in the northern Rocky Mountains. Animals that indulge in this tart–tasting treat leave typically reddish scats, common on this trail. Once, I saw the bushes shake, betraying a bear sneaking off. Upon arrival at the district station, the warden reported that he had ridden the trail a few hours earlier and met the biggest grizzly in his life. Another time, a lone hiker admonished me to be careful. A ways back on the trail, she had passed by a huge bear watching her from the shadows.

Yet, I continued to go there over many years, usually alone, although the spectre of bumping into the feared brute preyed on my mind. Traversing this neck of the woods, I habitually made a lot of noise, talking aloud to myself and hitting rocks and trees with a stick, even though I hated to disturb the quiet ambience of the virgin woods.

In my greenhorn years, I had been much bolder, not even bothering with the hiking trails. To explore the shores of a glacier lake shimmering in the valley below the park highway, I just descended cross–country through trackless forest. Or, to reach the alpine slopes above Bow Summit, where golden eagles were soaring, I simply parked the car by the side of the road and threaded my way up on foot through the tangled sub–alpine.

"If you keep on going to places like that, one day you'll meet a grizzly that knocks the head off your body!" This was the cheerless advice I received at the district warden station where I stopped to inquire about the occurrence of wolves and other wildlife. Eventually, coming to my senses, I gave up on solitary bushwhacking and stuck mostly to the established trails. Even this was discouraged by some of the locals. The grizzly was deeply feared and stories abounded of people who had been wounded or forced to climb trees in Banff National Park.

My very first encounter with the "monarch of the mountains" took place on the Egypt Lake Meadows. Despite all scary foreboding, it turned out to be an enchanting event. Ambling about in a flower–studded alpine draw, a sow and her three half–grown cubs were the picture of idyllic, wild beauty. Looking through the glasses at the happy family, grubbing in the soil, I was impressed by the awesome width of mama's fore arm. It was easy to imagine that a blow from that paw could break the neck of a moose, or send a person flying through the air!

Yet, this intimate and enjoyable first sighting could have easily ended very differently. When I spotted the bears, they were a safe distance farther up in a narrow gully transected by the trail. And they had not seen me. However, if chance had placed them right below the point where I had come over the crest, the encounter would have been abrupt and far more risky. Later, when I reported my observation to the district warden, he asked me to describe the sow's colour, which was brown with darker legs and blond highlights on the diagnostic shoulder hump. According to him, I had been lucky. This same bear, cubs in tow, had recently chased a guy on horse back and forced a party of fishermen to take refuge in a tree!

~

The grizzly used to go by the imaginative scientific label *Ursus horribilis*, but this was toned down to the prosaic *Ursus arctos* after the international community of zoologists agreed that the American grizzly was the same species as the brown bears of Europe and Asia. Wide–ranging and adaptable, they are true omnivores with a seasonal and varied menu that includes vegetation, roots, insects, fish, and any mammal they can catch or pick up as carrion. So why not humans, alive or dead?

Formerly, the predatory, man–killing grizzly terrorized the native peoples of the American West, but the guns of the European invaders forced it into retreat. In the Old World as well as in North America, mankind has waged centuries of war on the great bears, wiping them out in much of their former range. The grizzlies that survived the slaughter are less hostile. Yet, you never know. The word unpredictable fits the grizzly just as well as the sneaky black bear.

In the national parks, the reclusive monarch does not converge on campgrounds as readily as its more common cousin, but once a grizzly gets into the habit of approaching people for the purpose of scavenging, things may quickly go from bad to worse. In Montana's Glacier Park, some years ago, several campers were pulled out of their sleeping bags and mauled to death. A less deadly but similarly terrifying incident happened at Lake Louise in Alberta in September of 1995. Around 3:00 a.m. a grizzly tore open three tents and bit or clawed six foreign tourists who screamed and fought back in the dark. All received cuts and puncture wounds but recovered in hospital.

The aftermath revealed some embarrassing but understandable mistakes on the part of the government officials involved. After park wardens set up baited traps in the area, they caught a skinny sow and her yearling cub. Both were destroyed. As evidenced by their numbered ear tags, this same pair had been trapped earlier on a garbage dump in British Columbia and subsequently released near Banff National Park. At the time of their move, both mother and cub had been in good condition, but when recaptured later at Lake Louise they proved to be seriously malnourished. Apparently, the pair had gone through a tough time to find sufficient sustenance in their new surroundings, proving that the translocation of troublesome bears can indeed have negative repercussions for the animals themselves.

Adding more ironic detail to this tragedy was the fact that the Banff wardens had actually shot the wrong pair! This came to light after scientists compared the DNA from the skinny sow to hairs found in the ripped–up tents. These hairs had come from another mother grizzly and cub which were trapped later. Unaware of the fact that this second duo were the real culprits responsible for the campground rampage, the wardens had given the pair a free ride to British Columbia! All this might have been hilariously funny if the consequences had not been so bad for people and bears alike.

During the same summer when this sad comedy of errors occurred, grizzlies killed two people and wounded seven others in Alberta and British Columbia. The dead were a pair of hunters who had been butchering an elk they had shot. The bear had zeroed in on the smell of blood and killed both men before they had a chance to grab their rifles left standing by a tree.

Among the wounded in the other incidents was a moose hunter taking a break to wash and cool his feet in a beaverpond. He was stalked and bowled over by a grizzly before he could reach his gun. With his hunting knife, he stabbed the bear repeatedly in the throat. The animal pulled back and was later found to have bled to death. Severely wounded, the guy managed to crawl to where he had placed his rifle and fired off three shots to summon his comrades. They probably thought that he had just bagged a trophy. Instead, they had to carry the poor guy to the road on a makeshift stretcher of poles and jackets. Interviewed by the media from his hospital bed, the horribly scarred man still showed a sense of humour. Since the attack occurred after the guy had taken off his boots, he figured that the grizzly must have been attracted by the smell of his sweaty socks!

Far from a laughing matter was an attack that took place just west of Lake Louise in Alberta. An American gunshop owner who had bagged a bighorn ram was charged and thrown down a gulley by a grizzly that rushed out of the bushes. The hunter's two professional guides fled, pursued by the bear, but a cool–blooded young woman in the party, who was an employee of the gunshop owner, stopped the animal cold with four well–aimed shots.

Roughly similar altercations between hunters and bears were reported each year between 1994 and 1998. In some cases, the men had come upon a sow with cubs, but in the main the bears had been attracted by the fresh carcasses of game animals that had been shot. However, during the same span of five years, grizzlies also killed or wounded at least five unarmed people who just happened to be in the wrong place at the wrong time. In Yukon's Kluane National Park, in July of 1996, a man and his wife met a young male grizzly on the Slim Valley trail. Giving him the right of way, the couple slowly backed off, unstrapping and dropping their backpacks. The bear ignored the packs and seized the woman. When the husband tried to defend her, he was knocked down and the bear renewed his attacks on his initial victim. After some time, the desperate man went for help, an hour's walk away. Park wardens returned by helicopter, shot the bear and recovered the partly devoured body of the woman.

A few years earlier, a very similar fatality occurred in a backcountry campground on the trail to the famous Tonquin Valley in Jasper Park. All attempts by a British couple to withdraw from a young male grizzly and find safety in a tree proved futile. The woman was killed after a terrifying and hopeless ordeal.

Why did these bears attack? According to available information, it was neither a question of defending cubs nor surprise at close range. Given the remote location of the last two fatalities, it is probable that the young boars had no previous experience with people, either as adversaries or as prey. They probably did what comes naturally to their carnivorous kind and considered their victim as prey.

Although grizzlies spend most of their time grazing like docile cattle, their killer instinct may be aroused by the smell or sight of vulnerable mammals, be they elk, moose, caribou, or even another bear. In season, grizzlies search actively for the calves of hoofed mammals. They also move mountains of dirt to excavate marmots and ground squirrels, pouncing on them with the agility of a cat. Fond of flesh as they are, why should they make an exception of humans? Dragging a person out of a tent could well be as natural to this monstrous monarch as digging up a black bear that has just begun its hibernation. At least two of such gruesome mortalities have been documented by biologists, one in Alberta, the other in Montana. But who knows how many bloody and fatal brawls between these two hairy behemoths actually take place in the trackless forests of North America?

~

In the final analysis, neither fact nor fiction about grizzlies tends to bolster the backpacker's confidence on wilderness trails. Some shrug off useless worry and argue that the chance of meeting the wrong bear is infinitely slimmer than the risk of getting a traffic accident on the long drive to the mountain parks. This remains to be proven statistically. Nevertheless, given the great number of hikers who use the trails today, it is in fact nothing short of amazing how rare bear–human encounters actually are. From year to year, many summer months may go by without any shocking news reports. Considering that the number of grizzlies in the vast wilderness of western Canada runs into the thousands, the majority seems very well–behaved indeed.

A good question is whether or not mishaps are on the rise as compared to earlier decades. The answer is more complicated than it may seem since the relative size of the population is unknown. For instance, are there more grizzlies in Jasper now than two decades ago? And have they become less shy than they used to be? Bear behavior may change over time by continued protection inside the parks. Moreover, grizzlies are not hunted as hard as they once were on adjacent provincial lands. Spare the rod, spoil the bear, some say.

There is an interesting parallel with the Scandinavian countries where bears used to be scarce and extremely shy, and where protective measures have resulted in a spectacular comeback. The Swedish brown bear population has now surpassed one thousand. Experts have long warned that similar increases in Finland will eventually lead to human fatalities. And indeed, in 1998, a Finnish runner was killed by a sow with cub. In Scandinavia as well as far to the south in Croatia, the bruins have become more confident and are now frequently intruding upon farm yards and even into the outskirts of villages. This can only lead to more confrontations between man and beast.

How should we deal with bears that have lost their fear of humans? This is a topical question facing wildlife managers in North America as well as in Europe. Short of transplanting or killing bears, wardens in Jasper Park are trying to scare them off with rubber bullets. How about a dose of pepper spray? Would aggressive animals that are sprayed learn that it is wise to give us bipeds a wide and wary bend? Fact is that very few bears actually get hit with a squirt in the eye, and the smell of pepper alone does not deter bears. On the contrary! In Alaska, a guy who had sprayed the perimeter around his camp was flabbergasted to see grizzlies sniff and lick the substance from the ground and bushes! So much for the idea of spoiling a bear's appetite by soaking your clothes in the stuff. In fact, these four–legged gourmets might be all the more tempted by your personalized peppersteak!

There are those who argue that reliance on the spray may actually be counterproductive. Carrying any type of weapon might make people less careful and tempt them to drop precautionary habits in the false belief that they can save themselves in case of attack. Be this as it may, just having the can makes me feel less vulnerable than before. Whether or not I would ever be able to press the trigger under critical circumstances is quite another matter. If you come face to face with the monarch there may be little time to think or get your act together. The moment usually arrives when you least expect it!

~

One afternoon, after it had been raining cats and dogs for hours, all I had on my mind was to get back to the car and leave the cold and wet wilderness behind. As I rounded a turn in the trail, a huge, dark animal emerged from the woods ahead. A moose, I thought, trying to focus my near–sighted eyes. Wrong! Pulling the binoculars out of my jacket, I saw that it was a grizzly! He too had stopped, looking at me. Then he turned to face me. Expecting the worst, I took off my backpack and placed it in front by way of a shield, no matter how flimsy. Seconds passed like an eternity while the rain came down in torrents. Presently, the big boar moved on and vanished into the woods. Later, pacing out the distance that had separated us, I figured it was roughly sixty paces.

Another time, I got even closer to the monarch, so close in fact that I could not even focus. Like the previous experience, I was totally unprepared but for very different reasons. It was an early October morning. From the Lookout Hill overlooking the Snake Indian River, my buddy Peter and I had been observing a grizzly sow and her two–year–old twins. On a partly wooded island, the bears were digging for the tubers of *Hedisarum*, also called pea–vine vetch or licorice root, a plant that grows in abundance on montane gravel bars and is a grizzly favourite. We watched through binoculars as the bears quartered the uneven ground until they decided to leave. Mother in the lead, cubs following in line, they forded a channel of the river, traversed a semi–open meadow and filed out of sight into the woods.

Exited about our chance observation, Peter and I went back to our camp for a delayed breakfast. Afterward, we decided to investigate the island where the bears had been. Crossing the shallow river channel by stepping from stone to stone, we looked for tracks on shore and split up. Following a narrow game trail, packed down with the hooves of elk, I passed through a small poplar grove and ended up in a dry ditch, about three feet (1 m) deep and a few yards wide. Its soft muddy bottom constituted an ideal medium for tracking, and indeed, there were several sets of perfect prints of the bears.

"Hey, Peter, over here!" I shouted.

In the next instant, there came the clatter of animals running through the bushes on the island. Assuming that they were elk, I thought that my friend should have a better view than I, standing in the ditch, and I shouted again: "Can you see them?"

Then, the willows swayed apart and a huge grizzly burst into view, massive like a locomotive.

"A bear! Don't run!"

The animal came to a stop on the opposite bank, reared up on its hind legs and towered over me like a monstrous Sasquatch, its arms hanging by its side. The panicked footsteps of Peter sounded from the gravel bar, and I could not control myself either. I fled but not far. Behind me, I heard the grizzly jump down into the ditch. One of the poplar trees was easy to climb, but there was not enough time. Just as I reached the trunk, I turned and froze, binoculars slipping out of my hand. This was it. My day had come. In my mind flashed the image of bloody guts dropping onto the gravel. The headline in the paper read: Naturalist mauled by grizzly in Jasper Park....

In a state of shock, my eyes misting over, I vaguely saw the monster come down the game trail. A few feet away, she hesitated, huffed, then barrelled on by, splashing through the river channel. She was followed by a second grizzly, a big yearling cub, which came to a full stop at arm's length. It looked at me, snorted and ran off again, splashing through the water. The second cub repeated the actions of its sibling. It too looked at me for a moment before disappearing with the others.

When all was silent again, Peter found me still standing by the tree in a trance–like state. "Your face was as grey as your beard," he said afterward.

Upon arriving home in Edmonton and telling the story to Irma, I ended on a plaintive note that still makes her laugh every time the subject comes up: "I always wanted to see a grizzly up close. But I couldn't even focus."

The surprising thing about the encounter was that neither Peter nor I had the remotest idea that the bears would return to the island after we had seen the trio leave earlier that morning. Another revealing point was that the grizzlies had come toward the sound of my voice! This was contrary to the common assumption that talking and other loud noises help to ward off bears. Instead, the sow had come to investigate!

The third surprising fact was that neither of us had been able to suppress the natural impulse to flee, notwithstanding the fact that at the start of each and every backcountry trip my companions and I make a point of repeating our mantra: DO NOT RUN IF YOU SEE A BEAR! So much for human nature! Fortunately, I had not fled far. Later, I stepped out the distance as no more than six paces. One thing seemed certain, had I not left the trail just in time and frozen with fear, the grizzly would have knocked me down for sure....

~

As stated earlier, the luck I have had in avoiding serious bear trouble seems amazing. It fills me with gratitude as well as concern for the future. The dice may be rolled once too often. One day my number may be up. To be honest, the potential peril of running afoul of a mad bear preys on my mind, inhibiting the pleasure of solitary wanderings in the wilderness. Yet, I feel compelled to return. The hint of danger, the ancient fear of the beast lurking between the trees, lusting for one's blood, is an intrinsic part of the spirit of the Canadian wilds. It is part of the mystique of the northwoods, which are so different from the tame forests in countries where all our competitors, all large predatory mammals, were exterminated long ago.

Nevertheless, if we think hard and realistically about the terrible chances we are taking by travelling alone in the wilderness, and how often tragedy has struck others, is it not time to come to our senses and stop going? Bear experts such as Stephen Herrero, all too aware of the risks, strongly advise people to take every precautionary measure. One should not hike alone, certainly not walk silently through the woods, nor leave one's backpack with food standing by the tent. Instead, one should make lots of noise and travel in groups. Born out by research, bear attacks are usually aimed at people alone or in small parties, never at groups of over half a dozen.

Perhaps, as social creatures, we should take a leaf from our predecessors, the stone–age savages who survived over the millennia in the North American wilderness. Instead of sneaking about on their own, they operated in family bands. Similarly, African aboriginals, from the tiny Hottentot bushmen to the tall Masai warriors, live in tribes and they are armed with long spears that can be braced, butts stuck into the ground, to form a defensive shield even a lion would hesitate to charge. Like aboriginal people, our closest humanoid relatives, such as chimpanzees and baboons, always stick together for safety's sake.

Why then do so many modern–day North Americans choose solitary travel? Clearly, because it has its own set of rewards. Alone, not distracted by the chatter and concerns of the crowd, I experience the intrinsic atmosphere of the wilderness to the fullest. I no longer need the group for effective hunting and food gathering reasons. I can bring all I need. Nor am I depriving myself of social contact for weeks or months. All I want is to be free for a few days, away from the irritations of the city, away from the tyranny of the masses. Moreover, as far as wildlife observations go, the lone traveller sees more than a noisy group.

In my view, the best of both worlds is to have just one other companion, someone of compatible character and complementary skills. The benefits and joys of good company are priceless, in social interaction, emotional support, and badly needed help in case something goes seriously wrong. Translated, a Dutch saying goes: "Shared fun is double the fun, shared trouble is half the trouble." Moreover, according to a common maxim, two eyes see more than one. Problem is not everyone wants to or is able to visit the wilds as often as I do. So my choice was reduced to the simple question of not going at all or going alone.

CHAPTER 12

CAMPING BY THIRTY BELOW

Mankind's greatest invention is supposed to be the wheel. Be that as it may, I for one would nominate the airtight stove. On snowy winter trails, wheels would not get you very far, but few of life's luxuries beat the all–embracing warmth enjoyed by the foot traveller, who, coming in from the cold, sits down in front of a hot stove. By comparison, an open campfire, with Orion glittering overhead, may have its charms, and a well–organized bivouac certainly enables the experienced outdoors person to sleep under the stars and survive very low temperatures. However, once the flames die down and the coals turn to ashes, the frost will sneak its icicle fingers down your spine with a vengeance. No matter how much wood was burned, its heat is lost to outer space. Moreover, the frigid wind can bedevil what little shelter you managed to find. Your back freezes, while your face is scorched and your teary eyes sting with smoke.

Even our early predecessors, the nomadic tribes who roamed the western mountains, sought to retreat from the weather no matter how hardy they may have been compared to thin–skinned babes like us. Archaeologists claim that the Pocahontas Cave in Jasper's lower Athabasca Valley was inhabited at least as far back as six thousand years ago. The smooth face of a slab of limestone above the cave's hillside entrance still shows the stain of an open hand, printed long ago in red ochre paint. Arching a dozen paces back into the mountain, the vaulted cavern not only afforded protection from the occasional enemy, either animal or human, but mainly from the relentless wind. If you visit the cave today and use your imagination, you can still picture the members of the tribe relaxing on the animal hides that cover the gravel floor, while a blazing fire in the entrance radiates its warmth back inside. Women and children are happily chatting away, while the men folk boast about the day's hunt. Those who cornered the moose or sneaked up on the beaver are the heroes of the hour.

Today, the Pocahontas Cave stands empty. Year by year, more sand and grit fall from the eroding ceiling, covering the pellets of mountain sheep that litter the floor. A dusty corner is marked with the tracks and scat–burying scrapes of their enemy, the mountain lion.

Elsewhere in the northwoods, in the absence of suitable rock caves, aboriginals beat the winter wind in huts and tepees constructed from a frame of poles and covered with skins, birchbark or interwoven branches. In the arctic barrens, the most hostile of environments for humans, the Inuit built igloos from blocks of wind–packed snow. Mountain tribes might excavate a temporary snow cave into a drift. All of these crude but effective shelters needed an opening near the top where the smoke of fire or oil lamp was allowed to escape. Of course, a lot of hot air could be lost the same way.

While we may feel superior to primitive savages, we should not forget that they have proven themselves very capable and skilled at survival under adverse conditions. Subordinated to the ecosystem, they could sustain their lifestyle over the millennia. By comparison, it remains to be seen how long our greedy rush to rearrange the earth can last.

Given our society's emphasis on personal comfort, it is all the more notable that backcountry hiking and camping has become so popular. Perhaps it derives from an unconscious desire to reset our inner compass, to obtain a fresh perspective on our humanity and our place in nature. Furthermore, the mountain hiker gets to see remote country that is otherwise inaccessible. But why make it tougher than necessary? Why sleep in the open, especially in winter? I initially thought that my choice was limited to no tent at all or a pup tent. The lightweight nylon kept out the wind, but became coated with frost on the inside due to condensation of exhaled moisture. Little did I know that there was a much better alternative, a tent big enough to place a wood stove inside. This logical combination, unfamiliar to me in my greenhorn years, has been commonplace among hunters and trappers over generations. Seeing such a setup in the woods became an eye–opener. One day, out of curiosity, I followed an odd, rounded trough in the snow leading from the shoulder of a bush road into the trees. After some distance, it dead–ended at the trapping camp of a couple of native Indians. Their spacious tent, made of heavy canvas, could easily accommodate two beds, two chairs and a small box–type table. Most importantly, it was high enough in the centre to allow a tall guy to walk in without bumping his head. And in a corner stood a flat–topped barrel stove! Its chimney pipe protruded outside through an opening in the canvas protected with a collar of asbestos fibre. The absent but obviously resourceful owners had dragged their equipment into place in a makeshift toboggan, the upside–down engine hood of a car! Later, I found out that such tent and stove combinations are for sale in any traditional camping store.

~

When I suggested to Peter that we purchase an outfit, he immediately warmed to the idea, offering to go fifty–fifty on the costs. Over the years, we had made many backcountry trips together, including riding bicycles into the Willmore Wilderness at a time when the term mountain bike had not yet been invented. Now, in his fifties, Peter no longer cared for the many negative sides of the summer season, such as heat, rain, bugs, mud, and above all bears. He preferred the clean joys of snow and cold. Moreover, the job of setting up a roomy and warm tent was right up his alley. He liked to occupy himself creatively to organize the neatest camp possible, well–stocked with fire wood.

Our usual destination for winter trips was the semi–open valley between the confluence of the Athabasca and Snake Indian Rivers in Jasper National Park. Its alluvial bottomlands are characterized by a mosaic of montane meadows, mixed woods, and grassy slopes, which are a major wintering ground for elk and mountain sheep, as well as wolves. The shortest route to get there was by crossing the Athabasca River, which was close to the highway. If the water was open, we did so by canoe, but access was easiest after a prolonged spell of extreme cold when the ice of Jasper Lake became reasonably safe to walk on. This would also facilitate the transport of our heavy camping gear, which we intended to drag into place by employing another classical Canadian combination, toboggan and snow shoes.

One cold day in mid February, after obtaining the necessary camping permit, we parked the car by the side of the highway, strapped on our webs and set out. The snow was deep and fluffy. We sank almost to our knees, and trail breaking proved slow and tiring. Taking turns, the guy in front packed a base that made pulling the heavy toboggan a little less hard. We often joined forces, especially in tough spots. Before reaching the frozen lake, we had to negotiate an inlet of open, spring–fed water, which took some doing. But once across, we proceeded without a hitch to the opposite shore. Entering the woods, we found the snow deep again and our job became even tougher. The total distance we had to go was little more than three miles (4.8 km), but it took a good portion of the afternoon before we arrived at the neck of the woods where we had planned to set up camp. We were quite exhausted, but there was not much daylight left and we set to work at once.

The tent came without poles. We were supposed to obtain these from the forest. Arrow–straight and slender, the lodgepole pine is named for an obvious reason! Not far away, we sawed off two dead trunks of about four inches (10 cm) thick at the base, which we cut to a length of about nine feet (3 m). Stood up in the shape of an X and tied together at the crossover point, they provided support for one end of the horizontal ridge pole. Its other end rested at the required height on a branch of a big spruce. The gabled roof of the tent was hung onto the ridge pole with a series of knotted loops. The sloping halves of the roof were then stretched out at the correct angle by guy ropes fastened to metal stakes driven into the frozen ground. The vertical side walls, about three feet high (0.9 m), were studded with short poles set under the ropes.

The tent also came without floor. Our next big job was to scrape away the snow inside, using the webs for shovels. Here and there, the bumpy ground had to be cleared of a stump or small bush. Next, we cut armfuls of spruce bows and piled them on the floor as a platform for our foam mattresses. A few handy blocks and planks, salvaged from the ruins of a nearby cabin, served as seats and a makeshift table.

Finally, we were ready for the great moment of moving in the stove. One guy assembled the four sections of chimney pipe and stuck the top through the opening in the roof, while the other collected some fire wood. All it took was a fist–full of tinder dry branches and a match to start the fire. Adding thicker pieces, we soon had the flames roaring through the chimney and the sheet metal box began to radiate a near–tropical heat. We took off our jackets and hung damp clothing to dry. We then filled a pot with snow, placed it on the top of the stove, and busied ourselves with the preparation of a simple meal. Our winter standby was a rich mixture of brown beans and corned beef. By now, dusk had descended. In the intimate circle of candle light, we looked around us with contentment, a rosy shine on our ruddy faces.

"This is just delicious!" sighed Peter: "Such warmth, heating my old bones! This way I never feel at home."

Little wonder! The thermometer in a corner of the tent had climbed to 90° Farenheit (32.2° C). Warmed through and through, we could step outside without hat or gloves to look at the stars above. The sky was clear, the wind calm, and the outside temperature had already dropped to near zero (-18° C). It was going to be a very cold night indeed. Shivering, we returned to our abode. Under the black canopy of spruce, illuminated from the inside by the candles, the tent was a welcome womb of warmth and security.

~

One luxury of which we got plenty was sleep. Twelve hours in the sack became the routine, especially in the dead of winter when the sun set well before five o'clock. Tired as one is after tramping around in the cold, falling asleep was seldom a problem, but I woke often and lay waiting for dawn, twisting and turning on the hard and uneven ground. Later, we constructed more comfortable beds from a sheet of plywood, raised up on blocks of wood.

Between evening and early morning, the difference in air temperature in the tent could be as much as 120° Fahrenheit (49° C)! For, even though the little stove worked miracles with a few chunks of wood, it did not hold much fuel, and the canvas roof of the tent provided little insulation. Our tropical micro–climate was here one minute, gone the next. An hour or so after the flames died down, the inside temperature approached that of the outside. The coldest morning we experienced was 32 degrees below (–36° C)! Starting the stove was the very first priority! We always kept plenty of kindling handy. The best was nature's own: the dead twigs of spruce. A matter of minutes after the first guy got up, the snap and crackle of the flames would invite the other to stir. Soon, the hoar frost that had collected on the canvas above our beds began to melt and drip.

Our next concern was weather conditions outside. It was by no means always cold. On the contrary, in any month a blustery Chinook might blow up from the west and bring in relatively warm air from the distant Pacific, which could raise the temperature to well above freezing. Occasionally, even in mid winter, it rained, leading to all sorts of problems, such as slippery trails and a hard crust on the snow after the frost returned. A sustained melt–down could cause massive flooding of the lake. Many times we took quite a risk negotiating rotting ice. For extra security, we had devised a crude pair of skis consisting of seven–foot (2 m) boards fitted with a simple strap of leather to hold our boots in place. Dragging a long pole, we stopped often to test the ice ahead and find our way around open water holes. The lake was actually no more than a shallow widening of the treacherous Athabasca River and the current could erode the ice from underneath any place, any time. Problem was that one never knew exactly where the most dangerous spot was since the main channel could wander erratically, scouring out a new course. Carrying heavy packs and shuffling along on our clumsy two–by–fours, we must have made a pitiful sight, especially when the Chinook blew with hurricane force, driving sand and grit across the polished ice. Clouds of freeze–dried silt billowed down the valley, obscuring the view like fog and dusting each blade of grass and twig for miles downwind. These were the negative aspects of our chosen paradise! On days like these, I found little beauty in nature, and the life–threatening perils of crossing the lake made my throat dry with fear. There have been times when I, travelling alone and buffeted by a blustery gale, took off my pack and went down on hands and knees to negotiate a critical–looking stretch.

One February, when we had ventured our way in over the ice, we found the lake a vast sea of overflow water on our return two days later. Peter, always forcing himself to be tougher than he was at heart, insisted that we proceed as usual. Carrying a short pole to test the ice underneath the water, he pushed on stubbornly while I stayed behind, refusing to follow and fearing for his life. There were bound to be open leads, impossible to see under the ankle–deep overflow.

In moments of crisis such as this, we should take a leaf from the animals. When the going gets tough, the meek take command, not the heroes. The elk herd is led by a suspicious old cow, not the giant bull. Similarly, a band of bighorn sheep, never far from disaster, is guided across the precipice by an experienced ewe. With her sharp horns, she rebuts the impatient young ram trying to get by grandma.

If we were unable to cross the lake, there was a safe, albeit time–consuming, alternative: a fifteen mile (24 km) hike around to the nearest highway bridge. Reluctantly, Peter gave in. As it so happened, not far away a crew was working on the rails and they were just ready to leave. Finding room for us in their buggy, they dropped us off near the highway bridge. Moreover, the first truck to come by responded to my upturned thumb and gave us a lift to our car!

~

Henceforth, anxious to reduce the risks we were taking, we crossed the Athabasca by canoe below the outlet of the lake. Here, a swift and narrow stretch of river stayed open for most of the winter due to the inflow of relatively warm springs. Each time though, we had to carry the eighty pound (36 kg) craft through the bush from the highway to the water, an arduous task. We eventually replaced the canoe with a much lighter, inflatable dinghy. However, this did not mean that our problems were over. On the contrary! As before, a change in weather could spell trouble and block our return. During one early winter trip, after we had come in across the water, a blizzard plunged the region into the deep freeze. Two days later, we found the lake frozen except for the main channel. It was full of ice floes that were piling up into the outlet. The river was clogged but for a narrow funnel of very fast water in the centre. Dropping the raft into that channel would have been suicidal. The current would force us against the sharp edge of the ice, upsetting the craft and sucking us under. Yet, to my horror, my intrepid companion wanted to give it a try.

"Not on your life!" I protested, insisting that we make the long trek around. In the meantime, the temperature had moderated, the blizzard abated, and the sun had come out. Moreover, the warden at Snaring Station, eight miles (12 km) away, was at home and kind enough to give us a ride back to our parked car.

There were other dangers for us in store. The dinghy, though easy to inflate, came with a built–in hazard of which we were unaware. Its synthetic vinyl became weakened by repeated exposure to below freezing temperatures and sunlight. After a few winters of use, the seam of one of the two airtight compartment suddenly blew in the middle of the river! Fortunately, I was alone and managed to remain afloat, sitting in a bathtub that was very hard to handle. Reaching the safety of the shallows, I felt like the pope, ready to kiss the earth. Flinging myself headlong onto the gravel, I split the seam of my pants! An audience might have found this hilariously funny, but I was neither amused nor discouraged. I bought a second dinghy, which served us well until it too blew in the middle of the river! Again I was alone and managed to reach shore, cold and wet, thanking the Lord on my knees. Had there been two persons in the craft, we would have sunk.

There is a Dutch saying to the effect that even an ass does not make the same mistake twice. Yet, I went ahead and bought a third dinghy! This one, however, was made of old–fashioned rubber. It was light and small, just big enough for one person and a backpack. It provided several years of safe service until it was superseded by a beat–up, fiberglass canoe, acquired by Brian. The warden service allowed us to stash it in the shoreline woods after each trip, saving us the hassle of transport to and from home. Now all we had to worry about was ice. This could be a problem during early winter before the lake had frozen solid and the outlet was awash with floes. However, it seldom stopped us from finding our way across between the leads.

A greater challenge was solid shore ice if it extended well into the swift water. We could break it up before launching the canoe, but a border of fixed ice along the opposite shore made a landing hazardous or impossible. Once, travelling alone, my day was made by a herd of elk. They had crossed the river a matter of hours before, their sharp hooves cutting a swath through the shore ice wide enough to beach the canoe! After I landed, tracks revealed that my real benefactors had been the wolves. They had chased the elk out of the woods and forced them to flee across the river on the very place where I needed to row over!

Since conditions for each trip were different and our camp was quite isolated from the outside world, it was natural for us to be preoccupied with the weather outlook. This was all the more so if I went by myself. Useless worry was part of my mental ballast. Before leaving, I checked the official forecast, no matter how often it had proven to be unreliable. An ominous outlook could make me cancel my plans. But when I went with Peter, who had a demanding job schedule to consider, we never changed a date, blizzard warning or not. And often, the dire predictions were wrong. By the time we reached the mountains, the skies had cleared. To make up for the bad times, some days were just perfect, especially when the winds abated. Then, no matter what the temperature, and with fresh snow on the ground, it truly felt great to be there.

~

We spent the day roaming around, exploring the surroundings. The Devona Flats included rough meadows, thickets of wolf willow and mixed woods of spruce, pine, and a few poplar. A ridge of grassy hills rising over the east end of Jasper Lake were called Ram Pasture. They were the exclusive winter domain of the old boys band, a dozen or two Rocky Mountain bighorns. The ewes, with their lambs and yearlings, spent the winter months on the steep, south–facing canyon walls along the Snake Indian River. To the west, at the foot of the DeSmet Range, lay the Triangle Marsh, a vast, fan–shaped basin of beaver ponds and moose country, drained by Cabin Creek which bubbled up from sulphurous springs that flowed all year. The Triangle Marsh was connected to the river flats by a game trail, used mainly by elk. Their ancestral path followed the way of least resistance through the wooded hills and descended onto the flats right by our camp. A mile or two to the north lay another set of beaver ponds and small muskegs. It was fed by Cavanagh Creek that emerged from a crack in the rocky base of the Bosche Range. For some distance, its relatively warm rivulet stayed open all winter. Here, schools of small brook trout stood in the weedy current, skittishly dodging the attacks of the odd kingfisher, mink, or otter.

Like all of the wild denizens of this varied and virgin wilderness, we were creatures of habit with regular hang–outs and favourite trails. As far as seeing wildlife was concerned, we soon learned that hiking far and wide was less productive than staying in one spot. For instance, instead of chasing after the wolves and following their tracks, it was much better to let them come to us. Our pride of place was the Lookout Hill, a partly open ridge, about one hundred feet (30 m) high, right behind our camp. Ascending it via a short but steep trail, we sat down on a makeshift stool and relaxed, overlooking our stomping ground. Through binoculars, we frequently scanned the meadows and the river flats. If any animals came into view, they were not disturbed by our presence and we could observe them at leisure. We went up the hill each day for an hour or so, first thing in the morning and again in the evening. Even though we usually saw little or nothing, if there was a self–satisfied smirk on our faces, it was not because we felt like the masters of all we surveyed, on the contrary. The animals had ownership rights that went back over the millennia and we were but privileged observers.

On the last day of each trip, which usually included two overnightings, we took the tent down and stored everything in a hidden spot between the trees. Returning a week or two later, we repeated the process, reliving the pleasure of making ourselves comfortable in the bush. However, one day we were in for a rude surprise. All of the flats were covered in ice strong enough to walk on! Apparently, the river had frozen to the bottom and the irrepressible waters had found their way out overtop, inundating all low–lying land in the shoreline woods and adjacent meadows including the spot where we had stashed the camping gear. Fortunately, when we cracked the ice, using a stout pole, we found that the water had drained away and that there was now an airspace underneath the ice. Knowing where to look, we found our axe and set to work recovering the other items, a job that required many hours of hacking. Overtaken by darkness, we were forced to sleep under the stars, while light snow was falling. The long night in the open made us all the more appreciative of the comfort of our canvas abode, after we had reassembled it on higher ground.

~

Our happy camping days ended abruptly, not on account of the logistics involved, but because we ran afoul of the regulations.

The problem started innocently enough, one day in late fall when Peter and I were sitting by our camp fire on a secluded spot by Willow Creek. I for one have been called the world's worst dressed camper, but I never thought that Peter and I could be mistaken for old poachers. Yet, this is what happened. A fly fisherman, whom we had met on the trail, reported us to the park headquarters. Next morning, at the crack of dawn, two horse–mounted wardens arrived in our camp expecting to capture us red–handed. At the time, random camping was still allowed in the park. So we had done nothing wrong. However, in some quarters, the suspicion that we were poaching cast–off elk antlers persisted for a very long time.

In 1987, after the regulations had tightened up, requiring backcountry travellers to camp at designated sites only, I again ran into trouble by tenting on the same secluded spot along Willow Creek. The day was cold and wet, and my senior companion had become unwell, unable to continue on the muddy trail. There seemed no need to explain this to the patrolling warden. All he asked was: "Do you have a backcountry permit, Mr. Dekker?" I did indeed, and after a few remarks about the rain, the warden continued on his way. A few days later, I was phoned by headquarters. On the line was someone from the law enforcement unit. The message was curt. I had been charged with illegal camping. Subsequently, to tighten the screws so to speak, this young and eager warden had looked at my activities at Devona. He claimed that there was room here for a whole series of charges. For, even though random camping was still allowed during winter, it was restricted to the three months after December 21 and only if there was more than two feet (60 cm) of snow on the ground! This was news to me and to the people at the trail office, who had been kind enough to mail out the necessary camping permits for my winter tenting at Devona.

Just before my day in court, the warden offered me a deal: "Why don't you plead guilty for now, and thereafter we start with a clean slate." To avoid further unpleasantness, I took his advice and was duly fined by the judge. Soon after, during a meeting at headquarters, the new assistant chief warden said: "Instead of harassing you, we should be helping you." And in support of my wildlife investigations at Devona, he offered the use of the district cabin, which brought the camping area to a happy end.

Along the Peace River in northwestern Alberta, in August of 1965, we saw bears every day and their tracks were common in sandy or muddy spots near our camp sites. In northern Saskatchewan, we paid a visit to Billy, a native trapper and fire lookout, who had some hilarious stories to tell.

The author in 1961 on the open shorelines of the Stewart River in Yukon. Below, a smooth stretch of the Hootalinqua River, also in Yukon, reflecting the vast and peaceful wilderness setting.

We caught fish for supper and cooked over an open fire. Wind-bound during rough weather, Irma felt comfortable in the shelter of the shoreline woods, even though we might be several hundred miles from the nearest road or human habitation.

Nearly exterminated by the fur trade during the early 1900s, the industrious beaver is again common all over the northwoods. Photo: Robert Gehlert.

The cabin which the pioneering conservationist and author Grey Owl shared with a family of beavers still stands on the shore of Ajawaan Lake in Prince Albert National Park, Saskatchewan. After a sweaty portage to Ajawaan, Irma had her hands full to keep the mosquitoes at bay.

The humorous Indian guide George with his dog Cisko on the Nisutlin River in the wilds of Yukon. Along the Hootalinqua River, the author received some dire advice from a native woman, who was cleaning and preparing a batch of salmon for the family's traditional winter food supply.

During a winter trip on the frozen Brazeau River in 1961, Bill wore a heavy parka made of the fur of the animal we hoped to find. However, at the time wolves were very scarce in the Rocky Mountain foothills. They are common today, but typically grey animals such as the one portrayed are in the minority. Photo: Hank Wong.

The Willow Creek district of Jasper National Park is characterized by a mosaic of montane meadows, a fine habitat for observing wildlife. However, the hiker usually has to be content with their tracks. At left, the signature of the grizzly. Below, the foot pads of wolf and fox.

This rare photo, taken in 1872, shows the fur trading post called Jasper House. Then, the slopes of Roche Ronde were mostly bare, a consequence of forest fires set by native tribes. Now, after a century of fire prevention, the mountains are densely forested.
Photo above: Public Archives of Canada.

A band of Rocky Mountain Bighorn sheep on Ram Pasture, a traditional

Wintering range in the Athabasca valley of Jasper National Park.

To protect themselves against wolf attack, mountain sheep retreat into steep terrain, whereas elk take advantage of the relative safety of deep water. In winter however, their vulnerability increases and they too may seek a defensive position on rock walls or the edge of a precipice.

View from the Lookout Hill in the Athabasca Valley. Here, the author spent a few hours each morning and evening to look for wildlife, particularly during winter when the open meadows attracted herds of grazing elk. From March to November, the chance of encountering bears in the surrounding forest was always present.

Peter by the winter tent. A small wood stove allowed us to stay warm on the coldest days.

In support of our wildlife observations, park authorities permitted us to use a backcountry cabin. To reach the remote study area, we had to cross the treacherous Athabasca River.

The majority of Jasper Park's elk frequent the lower Athabasca Vallay, which is transected by the highway. In this less than pristine environment, grazing elk find abundant food and relative safety from their predators, the twin requirements for survival of all prey species. Photo Martijn de Jonge.

Author with skull and antlers of an elk killed by wolves. It is the paradox of the greatest bulls that they are often first to fall victim to their arch enemies. Photo: Brian Genereux.

PART 111

THE INTERTWINED WILD

CHAPTER 13

A DYNASTY OF BLACK WOLVES

In the early dusk of a stormy November day, my arms full of firewood, I came around the corner of the tent just when a black wolf emerged from the hillside game trail. It walked right by without the merest glance in my direction. Astonished, I quickly deposited my load and grabbed the binoculars. By now, the animal was some distance off on the meadow. Soon it would disappear behind trees. Hoping to stall it, I imitated a howl, ever so softly. The wolf halted at once and turned its head. Then, it raised its muzzle and answered, its sweet voice trailing off in a gust of wind. The questioning expression on its face was not at all what one would expect of a predator of its reputation. This wolf looked just like a friendly dog, eager to please and be petted. After a minute or so, receiving no further response from me, the animal resumed its course and trotted off in the bouncy gait typical of its kind, long fur blowing in the breeze.

Although it looked fair–sized, this was a pup of the year, some seven months old. Guileless, it had shown no apparent fear, but who knows how long it had been standing in the shadows, watching me chopping wood? Perhaps, following its parents, the pup had come by the tent earlier, while I was fast asleep. The resident adults certainly were not scared off by my camp. As indicated by tracks, they had investigated the site in my absence. One animal had even uncovered the metal tent pegs that I had left stuck in the ground, ready for hook–up next time. Such nosiness seems an undesirable trait for this species. What if there had been a trap hidden under the snow? Curiosity could kill. Although protected inside Jasper National Park, these wolves were certainly not without enemies once they crossed the boundary, less than eight miles (12 km) away.

On the morning following the charming interchange with the black pup, I saw the rest of its family, the Devona pack, for the first time. Numerous other sightings were to follow over the next two decades, but this early meeting was unforgettable. The wind had died overnight and I was on the Lookout Hill at sunrise. Soon after I howled, a wolf came into view on the meadow below. To my great surprise, it was white as the snow it trotted on and it was followed by five other wolves. Two of these looked like pups, both black. Either one could have been the animal seen the day before. The white wolf was a pup too, judging by its small size and baby–face profile. Of the three adults, two were black and one light grey. Remaining silent, the multicoloured procession soon filed out of view below the wooded base of the hill. Presently, they reappeared on a open ridge south of the lookout, screened by trees. Obviously aware of me, they kept glancing my way. Yet, they showed no concern and laid down to rest.

When a brief howl came from the far side of the flats, I scanned the area and discovered a seventh wolf approaching, a big black one. He was limping, favouring one of his front legs. Perhaps he had been hurt by a trap but he had managed to extricate himself. Obviously, he was having trouble keeping up with the pack and his short howl seemed to say: "Wait for me!"

In response, as if it had pity on the crippled straggler, the white pup raised its muzzle and sang: "Here we are!" Before long, other wolves chimed in, and one by one the pack came into full cry. On this still morning, the sun just peaking over Roche Miette, the howl gained in volume and filled the valley with an uproarious clamour. When the chorus died down, single wolves, perhaps the pups, kept calling off and on, as if they were loath to end the fun.

After half an hour or so, I was chilled and went back down to camp in a happy mood. My ears resounded with wolf music, and in my mind I saw a video film, starting with the moment the pack had come into view. What a sight it had been! And they had tolerated my presence! This is exactly how I had wanted our relationship to be, based on mutual respect. The last thing I wanted was to cause their departure. The privilege of sharing the hills and the meadows with these rulers of the animal kingdom was my ultimate reward for venturing into the wintry wilds.

With the intention of leaving them alone for most of the day, I went for a hike down the valley. Returning by late afternoon, I walked back up the hill and glassed the meadows below and the open ridges of adjacent hills. Nothing. The landscape was as grand as before, but it seemed empty and desolate. My spirits sank. What is a wilderness without wolves? No more than scenery. Across the valley stood Roche Miette, as always monumental, but its splendour was aloof and alien to me, untouchable, inhuman. How can we, mere mortals, comprehend, let alone "conquer" a mountain? Creatures of flesh and blood, we cannot relate to stone. But it is easy to feel kinship with "brother wolf." Like us, his pulse quickens to other living things. A wolf has to kill to live, but we can choose to abstain from blood letting, content to observe other creatures that remain closer to nature. As watchers and curious naturalists, we can probe animal secrets for our enjoyment and intellectual enrichment. But it is a one–sided love affair. Wolves do not wait for us.

Could they still be within earshot, I wondered? It had been calm all day, unusual for this windy mountain front and ideal for long–distance communication. I howled once and did not have long to wait. Eerily rising over the rolling hills to the west, the pack's response swelled in volume like the onset of a symphony, amplified in a theatre of snowcapped ramparts. Though far away, their song was pregnant with power. They were the masters of this timeless domain. In a flash of insight, I understood one of the reasons why wolves howl. Compelled to join voices in a bewitching ritual, each of the group members adds to their collective impact for the specific purpose of broadcasting the pack's numerical strength to others.

Another reason why wolves are so vocal is quite the opposite. Instead of keeping rivals at bay, it is to contact and communicate with their comrades. Earlier that day, the crippled wolf, trailing his group, had provided a telling example. Over the years, I saw several other times that lone stragglers located their pack after vocalising.

One winter morning, I got a distant response to my howl from two pups high on the slopes above the Snake Indian canyon. I went there and presently called the pair out of the woods at very close range. Ears pricked up and amber eyes shining in their black faces, they sat looking at me with eager anticipation, their heads cocked at an engaging angle. All too soon they became suspicious of my still, kneeling hulk and sidled back into the trees. A cursory investigation of the area turned up the remains of a mule deer buck, killed by the pack. After consuming most of the meat, the adults must have departed again, unwittingly leaving the two pups behind, while they were still chewing on the bones.

The following morning, the orphans were still alone but their wait was almost over. As I walked up the hill, they howled spontaneously. A while later, their pleading calls received the answer they were aching to hear. The pack called from the woods across the river! With a jubilant outburst of relief, the pups raced down the open hillside and somehow managed to descend the hazardous canyon wall. After crossing the frozen Snake Indian, they found their way up the steep, wooded bank on the opposite side, for presently I heard the excited sounds of their welcome. A lost pup's happiness in being reunited with parents and siblings can easily be imagined by anyone familiar with domestic dogs. How emotional they are! And how socially dependent! A lone wolf is a body without spirit.

~

During winter, packs travel far and wide across their hunting territory and individuals easily become separated, especially in rough terrain or when the action gets hot. Others, like the pair of pups, may stay behind to clean up the leftovers of a kill. If howling does not bring results, the straggler may decide to stay where it is or go back home on its own. Prior to the 1990s, when the Devona area was far less often disturbed by people than it is today, the local pack maintained a traditional rendezvous site on or near the meadows. Here, summer and winter, lost members returned to be united with their loved ones. The longer the wait, the more anxious they become and the more often they howl.

One black pup, which I heard call at midnight and again at dawn, seemed desperately lonely. In the morning, when I howled from the lookout, the wolf approached from the far side of the meadows. Coming up the hill, muttering to itself, it made a beeline for me. Imagine its shock upon discovering that I was a phony! The morning was bitterly cold and breezy, and I soon went back down to the tent to warm up. Returning an hour or so later, I found the pup lying up on an adjacent hill. Spotting me, it rose, snow sticking to the matted hair of tail and flanks. It began to howl plaintively. Over the next half hour, at a rate of six times per minute, it called roughly 180 times. Before that, since early morning, the animal had already vocalized continuously, possibly as often as 500–800 times in total! Its call was peculiar for it contained a split–second pause, when its voice apparently reached a high note that was inaudible for my ears. Moreover, the pup occasionally changed to a much lower tone, as if its lonesome mood gave way to a more assertive one, perhaps triggered by my vocal treachery.

The fact that wolves have feelings is not as far–fetched as it may sound. Those of you who own and love dogs will understand. One day, I detected an abrupt mood change in a lone wolf, who responded to my call. I recognized him as an adult male belonging to the Devona pack. He was on his way to the rendezvous site when I intercepted and stalled him by howling, not from my usual lookout, but from a sheep trail high above the valley. After his answering howl, he waited patiently while I remained silent. He lay down and raised his snout at intervals, repeating his greeting. Receiving no response, he finally decided that I must be an interloper and a coward, afraid to accept his invitation. He stood up abruptly and barked a threat that sounded like: "Go to hell!" Without further delay, he turned and continued on his way.

Did my habit of imitating wolf howls disturb and hinder my wild friends? And if so, why did I keep on doing it? Because it was my only way to make contact. Moreover, I had taken a leaf from their book and played an open card. By calling from the lookout in morning and evening, always from the same place, I announced my presence. Had I remained silent, my wanderings in their domain might have been more unpredictable and perhaps more alarming, although they had other means at their disposal to check me out. I am sure the local pack knew all they cared to know. Proof of this was that they often ignored me, not bothering to answer. At worst, my calls were a nuisance, like a wrong number on the telephone, but they had the option of not picking up the receiver.

At best, or so I flatter myself, my howling might be useful to the pack in their absence. I was flying the flag for them, so to speak, holding the fort. Strange wolves, loath to reply or show themselves in this hostile territory, might quietly vacate the area upon hearing my call. Occasionally, I may even have played a helpful role in bringing separated family members together. One day, I found myself between a straggler and a pack of five that had passed by earlier in the morning. The two parties were several miles apart. Responding to my howl, the loner answered from far off to the south. Presently, the pack reported from the river woods a mile to the north. After some time, the lone wolf came running across the flats, heading in the pack's direction. On the river, he paused to call again and waited for their reply. When it came, he shot away like a arrow. Presently, all six came out onto the ice, loafing and playing in the warm sun of a March afternoon.

~

Another question I have often asked myself is this: do wolves that hear my pitiful imitation actually think that I am one of them? Or do they respond just because of their innate, irrepressible urge to give tongue? In Bob Dylan's words, the answer is blowin' in the wind. No doubt, the local adults know me as a harmless oddball, but pups and strangers seem to find me believable. Frankly, as I mentioned before, my howl is a far cry from the real thing. No matter how often I have heard wolf song, I am simply unable to do it justice. However, this does not seem to be a problem for them. Perhaps, as with people, the way you say things is more important than what you say. In fact, a perfect copy of the wrong howl might be less likely to get a response than a poor attempt that echoes the right feeling. For instance, a biology student who performed the best simulated cry I ever heard, never got an answer. Lacking experience with real wolves, she had picked up her howl from a recording and her rendition was so accurate that I at once knew which recording. It was a commercially available record featuring Dagwood, a male wolf owned by the late Douglas Pimlott, the renown Canadian pioneer of wolf research. The young lady obviously had an excellent musical ear. Problem was that wild wolves did not seem to care for her howl because it was the call of a dominant wolf, possibly an unhappy and belligerent captive at that. Who wants to talk to a frustrated bully with a strident voice? By contrast, my feeble lament may transmit a sentiment that does not instill fear. I deliberately try for a mellow, high–pitched whine, indicative of an insecure and lonesome animal, looking for support and friendship.

However, from listening to the wild ones, I know that not all loners sound sweet and gentle. Some respond with little more than a one–syllable grunt before approaching me directly. I remember one big, grizzled–black brute who had approached silently until he stood behind me in the trees. The moment I saw him, he turned away. If I had been a real wolf, I might have been in trouble!

On the whole, my vocalizations have proven to be surprisingly successful in getting responses from the wild, not only in terms of hearing wolves, but also observing them, which was my primary objective. If the pack showed itself on the flats below the hill, I had an opportunity to count their number and recognize individual animals.

~

How often have I actually seen wolves, you may ask? Over a period of twenty winters, from 1979 to 1998, I got lucky on 108 days. The actual sightings might involve either a single animal or a large pack. The total number of wolves seen in these twenty winters was 429. The average or mean group size was thus close to four. But this was not necessarily the whole of the Devona pack. When wolves go hunting they may split up and rejoin later on. In determining the numerical status of the pack, I went by the largest group seen each winter. In addition, on days when all I saw were tracks, I counted the number of individual spoors. In some years, the Devona pack was also observed in adjacent districts by park wardens. All together, we ended up with a reliable assessment of winter pack size. The number changed from year to year and varied from five to thirteen, except for one winter when I saw no more than two. What had happened? Had the pack been poisoned or shot on provincial land outside park boundaries? Fortunately, the following January a pair got together again at Devona and rejuvenated the dynasty. They may have been newcomers, dispersing from elsewhere, for when I saw them, their reaction to my howl seemed tentative and hesitant. Shrouded by snow flurries, the pair came trotting upriver quite far apart. They dawdled as if they were exploring the place, calling back and forth to each other.

As they drew near, I could see that the leader was large and grey with a dark plume on his tail. His mate was black and quite a bit smaller. Both passed out of view in the trees below. A little later I went down to look at their tracks. The male had come partway up the hill until he encountered my trail. There, against the snow coating a tree stump, he had sprinkled his yellow signature, and to show how big he was he had aimed as high as he could! Little did he know that, on that score, I could have outdone him anytime! Satisfied that there was no sign of a real wolf in the area, he and his mate retreated and went on to establish a territory. By the fall of that year, when I saw them again, they were accompanied by four pups.

Over twenty consecutive years, the mean or average size of the Devona pack was just under eight (7.9) members. This is practically the same as the mean of 7.8 reported by biologist Carolyn Callaghan from adjacent Banff National Park. Her sample was based on thirteen packs which she had tagged with radio–collars. For further comparison, a team of researchers led by renown wolfman David Mech found a mean of 6.7 in 112 packs in Denali National Park, Alaska.

Apart from their numerical size, how can one distinguish one pack from the other? As you may have guessed, the answer is easy. Since western wolves come in all shades from white to black, including tan, silver, and rufous, it is often possible to recognize individuals and their packs. This should be much more difficult in eastern North America where wolves are predominantly buffy–grey with very little variation. With their rusty ears, they look like oversized coyotes. Black animals are very rare in the East while common out West. Here are some figures.

In three large samples (254–498) of wolves killed in control actions in Alaska, British Columbia and northern Alberta, blacks made up a little over thirty percent. In Jasper park the proportion of melanistic animals is even higher and roughly half. It was fifty–five percent in a total of eighty wolves sighted by wardens during the 1940s, and forty–six percent of fifty–seven wolves seen by Ludwig Carbyn three decades later. Quite similarly, my summer sightings near the dens in Willow Creek, which totalled 132 individuals minus obvious duplications, included fifty–three percent black.

By comparison, my winter sightings at Devona show a much higher percentage. From 1979 to 1998, I saw 157 individual wolves in twenty packs. Of these, seventy–three percent were black! This is the highest figure ever reported from anywhere in North America. In 1992, when the Devona pack was at its largest, all of its thirteen members were black. Between 1992 and 1996 I did not see a single grey wolf in the area, and very few were reported by wardens from elsewhere in the park.

Where did I draw the line between black and grey, considering the various in–betweens? I categorized a wolf as grey if it was pale tan or yellowish on its throat, belly, and legs, with or without darker accents on head, tail or along the spine. Black wolves could vary from completely black to smoky–grey with black legs, tail, or ears. Interestingly, with increasing age, black wolves can fade and change colour quite dramatically. According to Monty Sloan of WolfPark, Indiana, captive animals that were dusky as pups have been known to turn silvery–white in a few years. In Yellowstone Park, where over half of the reintroduced Canadian wolves were black, at least one has turned white in a matter of two years. Its identity and age were known since this animal had been collared by park biologist Douglas Smith.

White wolves were rare at Devona. I saw only four. Just one of these, the pup mentioned above, was the colour of snow. The others looked creamy white.

Since wolves, like people, turn grey with age, the very high proportion of blacks in Jasper may indicate a young population. This is probably true. The oldest known wild wolf in North America was thirteen years of age, whereas captives have reached sixteen. The average life span of the Devona wolves is certainly much lower. Vulnerable to trappers and hunters on park boundaries, they are also frequent victims on the busy highway that transects the main Athabasca Valley. Warden Wes Bradford, who tabulates traffic mortalities among the park's wildlife, picks up one or two dead wolves each year, sometimes as many as half a dozen. Others become victims of trains. In hot pursuit of elk, wolves have been known to run into a locomotive they could see coming from a mile away. The list of casualties is expected to grow as the transportation corridor through the park gets busier.

~

The colour composition of the Devona dynasty could well change in favour of grey. In March of 1998, a pack of seven contained only three blacks. Earlier that winter, a smaller pack of five included two blacks and three greys. This was also the very first time I had seen two different packs on the meadows! Each included animals that could be recognized as individuals. One of the packs had been observed by wardens at Pocahontas, the other at Snaring. The fact that I had seen both packs from the Lookout Hill indicated that Devona was on the boundary between them or that their territories were overlapping. Perhaps, the alpha wolves of both families were related and had originally split off from the Devona dynasty to team up with grey newcomers. Establishing separate dens a dozen or so miles apart from each other, they had boosted the valley population to a new high. It would have been interesting to see the two packs interact, either chase each other or mingle on friendly terms. It did not come to pass.

Initially, the summer of 1998 looked very promising. As reported by warden Greg Slatter, the Pocahontas pack produced six pups. I found their tracks at Devona. Unfortunately, by December, the pack had declined again to four adults: two greys, one white and one black. They were the scruffiest bunch of wolves I have ever seen. All of them showed signs of mange, a potentially lethal affliction of the skin caused by mites and leading to loss of hair. Apparently, the pups had succumbed to the contagious disease, and by late winter several adults may have died as well. Moreover, there was no evidence that the Snaring pack had brought off a surviving litter either. In the first three months of 1999, I saw no wolves at all at Devona, and the largest group tracked was two.

In early January of 2000, the activity of ravens betrayed the presence of eight wolves, seven blacks and one the colour of silver. In November 2000, the local population had grown to a dozen. One of these was killed on the railway, and another became a lone pack follower. All of these twelve wolves were again black!

Whether or not the blacks will continue to dominate and perpetuate this dynasty into the new millennium is a question to which the answer is blowing in the breeze, until the next time I see my wild friends from the Lookout Hill.

CHAPTER 14

THE BIGHORN'S DILEMMA

Hiking quietly through the still woods, I could not escape the sharp ears of a hunting wolf. But all I saw of it was the flash of its tail as the animal turned around and vanished between the trees. The few times such a thrill has come my way, I felt empathy with its prey species. Although modern–day wolf lovers tend to view the once despised outlaw as a furry mollycoddle, a lupine companion to Bambi, the real wolf is undeniably one of nature's fiercest killers.

How do defenceless creatures such as the hoofed mammals, that lack our untouchable immunity, cope with their treacherous foe? The answer to this intriguing question was slow to emerge and rarely through direct observation of the chase. After more than three decades, all I got for my patient watching were no more than a few glimpses of the drama that takes place in the wilds every day. Sure, I have found the remains of animals killed by wolves. But I have never been witness to the kill. To learn about the interaction of the hunters and the hunted, I had to identify with the prey and share its fear. How often would a deer see wolves make a kill? How often would it be chased? We can only guess. Nevertheless, its trembling instincts are acute and guide its daily habits, for the threat of predation is ever–present.

To any animal, the need to eat is as basic as the need to avoid being eaten. Its chosen environment should provide plenty of food as well as reprieve from predators. In the wild, such a paradise is hard to find. Some places may offer safety but little in the way of fodder, and vice versa. Most everywhere, herbivores live on the razor's edge between security and a full belly. A good example is the Rocky Mountain bighorn sheep. This ungulate is perfectly adapted to stony highlands and its life–history is an illustration of how to cope with the wolf.

In Jasper, bighorns are the most numerous of the seven species of hoofed mammals native to the western mountains. A rough estimate of the park's sheep population, according to the warden service, is close to three thousand. Exact totals are difficult to come by in this rugged wilderness, but sheep occur in all suitable habitat, and as far as is known the herds have remained remarkably stable for decades. Their home ground includes grassy range as well as precipitous cliffs, the only place where these docile and defenceless vegetarians find sanctuary from their speedy and relentless enemy. The wolf's love for lamb chops has forced its supplier to retreat into the vertical, where natural selection—the survival of the fittest and quickest—has created a mountaineer par excellence. With their sturdy physique and concave, anti–slip hooves, sheep can out–climb any wolf. Coolheaded, standing on an outcropping of rock, the bighorn stares down his foe, just out of reach of its dripping fangs. But to find a good pasture is quite another matter.

During the short summer, when the radius of action for most wolves centres on their dens in the valleys, the sheep ascend to the greening tundra above timberline where predators are few and far between. While the rams are footloose and fancy–free, the ewes are charged with the responsibility of raising the next generation. Guided by an experienced granny, small bands of females, still accompanied by last year's surviving youngsters, follow age–old travel routes to the lambing grounds. In June, the pregnant ewes isolate themselves from the group and seek out hiding places among a fortress of boulders. Soon after the miracle of birth, the dapper lamb stands up on wobbly legs, and in a matter of hours it can keep up with mom, who rejoins the herd. In a time–tested custom, the sisterhood aids new mothers with collective baby–sitting services. Leaving her precocious baby to frolic with others in the nursery band, the ewe is free to roam the sparsely vegetated uplands where finding food is a matter of eating on the go, nibbling a leaf here and a flower there. From time to time, she returns to attend to her newborn. If she lies down to rest and chew her cud, her impatient lamb jumps up on her back and prods her with a tiny hoof: "Hey, mom! I'm hungry!" When she reluctantly stands up, the little guy butts her udder to make the milk flow.

The absence of trees in the alpine allows the sheep to spot a predator from afar. Their eyes are excellent. At the discovery of a wolf, the herd goes at once into panic–mode and runs for the nearest rock wall. Strange as it may seem, quite unlike the steadfast females of elk or moose who care for their calves individually, ewes do not defend their young. Living in a commune, they seem to show little maternal concern. In a crisis, individual survival is paramount, which makes good sense. Any ewe that is so foolhardy as to turn back to protect her lamb from a pack of wolves would be certain to die as well.

Coyotes or cougars elicit the same flight response as wolves. Bears are avoided only if they come too close. But an enemy from which there is no escape is the golden eagle. A sheep that tries to run away would present the swift bird with an opportunity to clamp its talons into the prey's back. Instead, recognizing her peril, the smart ewe stands her ground stubbornly, confronting the attacker, while sheltering her lamb under her belly. She does not always get the chance for such heroics, though. Swooping down unnoticed, an eagle might grab her baby before mother has any warning at all. One by one, casualties mount. Others succumb to inclement weather, and by fall less than half of the lambs are still alive. Based on my long–term counts at Devona, their number will decline further to around two or three for every ten ewes.

For sheep, and indeed for all other warm–blooded creatures, winter is the critical season that sets the final quota on their kind. Before the high country is covered with deep snow, the band descends to traditional ranges on lower ground that meet with their twin requirements of ample food supplies and relative safety from the enemy.

~

The Athabasca Valley features a number of sites where wintering sheep congregate. One of them is the south–facing slope above the Snake Indian canyon. Here, come October, several small bands get together, mainly ewes, yearlings, and lambs. The total differs from year to year, varying from thirty to seventy, but my average or mean count over two decades is around forty. For the next seven months, the sheep travel little, preserving energy and eking out a living on their restricted range that becomes more depleted as the days go by. The wind is their ally and their enemy. The Chinook keeps the steep ground bare, but a blizzard combined with low temperatures forces the sheep into the shelter of adjacent woods where snow lies deep. Scraping it away with their hooves, the animals have to work hard. Besides, between the trees, they are vulnerable to predators. By the time I find evidence of kills, there is not much left but hair, a few bone splinters, and part of the skull.

Weather permitting, the band leaves the oppressive woods and returns to the canyon slopes. If the blizzard has filled the hollows with deep drifts, the size of their pasture is reduced in acreage and the sheep have to make do with even less. But in the open, they can at least relax and chew their cud in relative peace, although they never drop their guard. As soon as a set of furry ears protrudes above the brow of the hill, the band dashes for safety. When I find them standing on the near vertical cliffs above the river, I know they have seen the enemy, which more often than not remains hidden from me.

In November, the herd's pastoral routine is interrupted by an annual event. It is mating time. The rams, which have spent the summer in separate bands, come down from the highlands to join the ewes. Sleek and fit after a season of rest, the big boys strut their stuff, showing off the size of their curved horns to each other and spoiling for a fight. As always, it is the weak who challenge the strong. Rearing up on his hind legs, the upstart invites his superior to go head to head. They slam their horns together with all the force they can muster. Dazed by the bone–shaking impact, the combatants pause for a moment, while the echoes of their collision reverberate along the canyon. The fight is about dominance over their competitors. The champion is first in line behind the ewes. Vying for their fickle favours, he follows his chosen mate and frequently smells her urine to test for receptiveness. Unwilling, she dodges his attentions and flees into the rocks. The suitors pursue, rebutting each other, until she stands and obliges. The act of mating is brief and frequent, and both sexes are promiscuous.

~

This rather dry description of the sheep's sexual mores may suffice from a zoological point of view but does not do the animals justice. To the human onlooker, the breeding rituals of the bighorn can involve individual affairs worthy of a comic opera. One December day I watched the shenanigans of two rams chasing different ewes. One of the suitors was a trophy–sized old boy who had set his heart on a slip of an ewe, a mere yearling. The other ram, a lightweight upstart with small horns, was chasing a mature ewe. Although neither of the females seemed in the mood for sex, the rams did not take no for an answer. The scene that followed made me chuckle, although I might well be guilty of reading too much into it.

The young ram shadowed every move of the skittish old ewe, who darted across the steep rimrock of the canyon wall. As soon as she paused and he caught up, he eagerly raised his front legs and placed them on her rump, ready to copulate, but each time she ran from under him and resumed her flight. At last, after having been chased back and forth many times, she faced a solid wall and could not, or would not, go on. He mounted her and had his way in a quick shuddering thrust. He then stood quietly behind her. A few minutes later, he left her alone. If they had been humans, I am sure she would have had grounds for charging him with rape! But here too, my interpretation may well be wrong!

By contrast, the randy old boy was not getting anywhere. Top–heavy with his massive horns, he was not manoeuvrable enough to chase the nubile ewe over rough terrain. She had found a precarious refuge on the lower wall of the canyon, a near perpendicular slope of grit and shale. In delicate balance, she remained immobile, while all he could do was look at her. He dare not move for fear of sliding down. Both stayed put, a dozen paces apart, while the other courtship chase went on in the cliffs above them. At one time, the young ram dislodged a boulder that ricochetted down, just missing the other couple, and splashed into the river below.

The young ram's final triumph seemed to stimulate the old guy into action. Slowly and deliberately, he descended obliquely toward the object of his desire. At every step, his hooves were loosening stones and grit. As he got close, she dodged and moved on a little ways, staying out of his panting reach. The ram persisted, and at one point the ewe seemed to have reached a dead end. While she stood stoically, he inched up to her and rested his heavy head on her shapely rump. Then, ever so gingerly, in slow motion, he began to raise his front legs up onto her back. But as soon as he brought the business portion of his anatomy forward to her posterior, his rear feet slid from under him. Jumping aside awkwardly, the ram just managed to save himself from plunging down the precipice. His impotent siege continued until I got too cold to watch further.

~

By the time winter begins in earnest, the sheep settle down to the sober business of survival. Their passion spent, the rams depart for their own traditional wintering range, the grassy hillsides above Jasper Lake. Before starting out, they gaze into the distance for a time, gathering courage. From the canyon, the upper slopes of the Ram Pasture can be seen, but to get there, the sheep have to pass through a series of wooded draws, far from the safety of cliffs. Once on their way, they press on. If they happen to meet me on the Lookout Hill, which is in between the two sheep ranges, they dash by at arm's length after a moment of hesitation. They apparently recognize me as the harmless biped they have often seen from across the river.

One by one, the battle–scarred warriors arrive at their traditional hang–out where they are to spend the next five to six months in each other's close company. Like the plains Indians of old, the ram with the biggest headgear is the chief. Their hierarchy involves subtle rituals that keep aggression to a minimum and allow group members a certain amount of freedom. According to Canadian sheep expert Valerius Geist, the rams live in a homosexual society that includes a lot of body contact. Huddling together, the subordinates rub shoulders with the dominant male, horning his face to absorb his scent produced by the preorbital gland under the eye. In this way, they acquire a common odour. Reaffirming their status, the rams salute each other by standing stiffly at attention, neck stretched out. Cocking their head, they display the width of their horns, like a badge of honour, without ever looking each other into the eye. Staring is a sign of aggression. Any subordinate that is so bold as to look directly at the dominant ram is put back in his place with a swift blow of horn on horn. When at rest, the sheep avoid eye contact by facing outward in the same or a divergent direction. This makes it next to impossible for their enemies to sneak up undetected. Moreover, the rams are ever vigilant and their bulging eyes allow the widest possible angle of view. They can almost see behind them.

In open terrain, the sheep are secure as long as they do not stray too far from their refuge. However, as the days go by, there is less and less to eat on the slopes closest to their cliff. Gradually, the rams are forced farther away from their safe–haven. The best grazing is where the spruce woods begin. If I see the band feeding between the trees, I know that they have been left in peace for some time. On the other hand, if all of them are standing on the cliff, they have had visitors! Tracks reveal that the pasture is checked out by wolves one or more times each week. An hour or so after the danger has passed, the sheep slowly edge away from their fortress and return to feeding. I can also tell if they have been chased hard and repeatedly. Then, they startle as soon as the top of my toque rises into view over the sloping terrain. If I climb higher, they recognize me and relax again, resuming whatever they were doing.

Despite the seesawing demands of their dilemma—how to fill their belly and save their hide—the rams hang on to their hazardous range. No matter how precarious their existence may seem to us, the records show convincingly that they are coping well. For weeks and months, their number stays exactly the same. Apparently, they rarely get caught. After all these years, I have found no more than two wolf kills. Both rams were intercepted halfway up the slopes. In one case, I came across the scene while the pack was still feeding. In the other, a pair of wolves was sleeping soundly on a full belly, a little distance away from the partly consumed carcass.

The rams' low mortality rate not only indicates that they can take care of themselves, it also illustrates the weakness in the strategy of their enemy. No matter how often the sheep escape downhill, heading for the same cliff every time, the wolves seem incapable of figuring out their problem. By contrast, a human hunter would easily outsmart these rams by approaching from below and cutting them off from their predictable refuge! In the old days, these rams would have been at the mercy of aboriginal tribes, who were deadly cooperative hunters. A couple of men would hide by the cliff, while others drove the band down toward them. Herein lies the difference between humans and the beast. No matter how fast the wolf may be and how sharp its senses, its animal brain is clearly incapable of comprehending the logical connection between action and counteraction. By the same token, this handicap is exactly why these natural predators have been able to coexist with their prey over the millennia instead of hunting them down to the last, as our kind has been known to do.

Another indication that the sheep are standing up to the enemy is that the band's size has remained quite stable over the years, averaging some twenty mature, full–curl rams each winter. Their number is primarily determined by the food resource, by the carrying capacity of their winter range which is exploited to its full extent. By mid May, when the sheep finally depart for their alpine summer retreat, there is not much palatable vegetation left. The grass is grazed right down to the ground. Often the sheep paw up the roots as well, leaving nothing but bare soil. Clearly, sheep numbers are limited here by food supply, not by wolves. However, it is undeniable that predators play a critical role as well. If it were not for the threat of wolf attack, the sheep population would be free to increase since nothing would stop them from expanding their winter range. However, in that case they might get in the way of other herbivores, such as the elk.

Competition for food between the two grazers is an additional limiting factor on the Ram Pasture. In most winters, especially if the woods are choked with deep snow, some two dozen elk bulls join the rams on the windblown escarpment above Jasper Lake. They contribute to overgrazing of the range, which may lead to damage that can take years to heal or become permanent.

Disturbed ground opens the door to noxious weeds such as the Russian thistle or tumbleweed. Originally an accidental import from Europe, this aggressive colonizer, driven by the wind, has rolled across the entire western plains. It has invaded the Athabasca Valley via the railway. From there, seed–carrying fragments were blown onto the Ram Pasture, sprouting wherever they found a patch of bare soil with sufficient moisture. The mature plant hardens to a round bundle of sharp thorns, avoided by all grazing animals.

Today, the prickly weeds are crowding out native plants on part of the pasture. Since 1994, when I first discovered this problem and alerted the warden service to it, the thistles have spread, leaving the bighorns even less room to forage in safety, thus aggravating their dilemma.

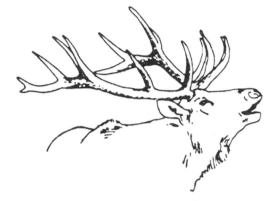

CHAPTER 15

PARADOX OF THE HAREM MASTER

In wildlife watching, memorable observations usually come your way when you least expect them, and they can make up for years of patient but fruitless effort. The first time I actually saw wolves attack elk happened one June afternoon when I had just set up camp and was walking down to the river to get a bucket of water. As I emerged from the woods, two black animals trotted by, screened by willows. With a start, thinking they were bears, I fumbled for the binoculars. There was just enough time to aim the glasses. They turned out to be wolves.

Leaving the bucket on the trail, I hurried up the adjacent Lookout Hill, hoping to get another look at the pair. As soon as I had gained enough altitude to look out over the flats, I halted, quite out of breath, and discovered the wolves on a semi–open island of the river. They had cornered a group of four elk, all bulls with large antlers, still in velvet. To steady the glasses, I sat down, expecting to see a great battle. But it turned out to be a dud.

Confronting and dwarfing their adversaries, the elk refused to run. If a wolf got too close, one of the bulls advanced, threatening to strike with his hooves, but he turned around just as quickly again if the other wolf sneaked up from behind. After a few minutes of tentative moves and countermoves, it became clear that the hostilities were at a stalemate. The elk began to nibble the vegetation, and the wolves ambled back and forth indecisively. Eventually, they disappeared between the bushes that covered most of the island.

The above chance observation illustrates the paradoxical aspect of this predator and prey relationship. Evidently, elk are quite capable of defending themselves. Yet, this large species of deer—also known by the Indian name of wapiti—represents a major portion of the food of wolves in Jasper. During their field studies, respectively conducted in the 1940s and 1970s, biologists Ian McTaggart–Cowan and Ludwig Carbyn calculated that elk made up over forty percent of prey consumed by wolves. This figure was based on the analysis of hair contained in several hundred wolf scats. Proportionally, the largest share of the elk sample had come from calves. This shows how predation works. Wolves kill the weakest individuals of their prey species, those least able to defend themselves.

The altercation I had witnessed, involving the four defiant bulls, was no more than a test of wits and fitness. No doubt, the two wolves would have been more persistent had they come across vulnerable prey such as a cow with a newborn calf. The river shore was the right place to search for these, and this was the right time of the year.

In June, pregnant elk cows withdraw to the seclusion of islands. The vegetation on the fertile riparian soils is lush, supplying abundant food, and the water breaks the animal's scent trail, which translates into refuge from predators. Safety is not guaranteed however. Wolves are in tune with the secret world of their prey. They know all about elk and getting wet is no problem for them. Entering the river without hesitation, the local wolves are like spaniels that swim the fastest and widest stream.

Tracks in shoreline mud attest to their frequent riverine patrols, but in this brushy habitat, the animals themselves are hard to observe, let alone in pursuit of prey. After all these years, I can still count the chases I have seen on my fingers. If I caught sight of the elk in full flight, their enemies usually remained hidden. Several times, when a small herd ran out of the woods and hurriedly entered the river, I waited with bated breath. One of the elk standing in the water had suffered a bleeding wound to her rump. Even then, no wolves showed up. Perhaps they had already made their kill in the woods.

Brian observed half a dozen elk, chased by wolves, swim across a wide channel of the Athabasca. In the water, the predators actually gained on their prey. The pursuit continued on the opposite bank, all too soon gone from view. Another time, Brian and I watched a panicked herd run up a hill. Suddenly, one of the cows turned back down, bleating anxiously. Apparently, she was calling her missing calf. By the time we discovered its bloody remains on the shore of a beaverpond, the killers had already left.

In a moment of crisis, the river can provide a refuge that makes the difference between life and death for elk. Their long legs allow them to make a stand in swift, belly–deep water where their assailants are swept off their feet. Once I watched a wolf swim up to an elk cow standing in a rapids. She coolly waded upstream out of her enemy's reach. The wolf, only its black head visible in the turbulence, managed to find a shallow spot. There, it braced itself for some ten minutes, while its companions waited on the shore. Presently, all of them gave up the siege. The cow stayed in the water for another twenty minutes, until she hesitantly and stiffly returned to land. Her fate might have been quite different had the wolves attacked from several sides. Instead, they had given up soon and departed in the same direction as the other elk had gone. The herd included several cows with calves. They had been first to flee across the river when the pack had suddenly sprinted out of the woods.

The toll that predators exact from the elk can be assessed by the low number of calves left with the herd later in the year. As part of my observations, I count and classify the elk seen from October to March. Overall, the average or mean number of calves is often less than twenty per one hundred cows and yearling females. Assuming that the majority of mature cows should have given birth that summer, where have their calves gone? No doubt, most of the missing ended up in the jaws of wolf, bear, or cougar.

Elk recruitment is significantly higher in places where predators are less common, such as near the Jasper townsite. There, as reported by warden Wes Bradford, calf percentages range between forty and fifty percent, more than double my figures for Devona. The total number of elk I see each winter, and the maximum size of the herds, are also very low compared to those that frequent the townsite and the highway corridor. The elk instinctively congregate in areas where they feel safest. This phenomenon is termed an anti–predator strategy.

The best time to count elk cows and calves is autumn, when the fast–growing youngsters are still small enough to be easily identified at a distance. And a distant look is usually all I get, because the local herd is extremely wary and easily spooked. As soon as they see me, one of the cows emits a squealing bark and all of them take off in a tight group. If pursued by wolves, the few remaining youngsters are protected in between the flailing hooves of the adults.

To anxious elk mothers, early warning is of vital importance. For that reason, they prefer open country, which is therefore another anti–predator strategy. Quite different from myopic woodland ungulates such as deer and moose, elk have sharp eyes. At the sight of danger, the herd usually heads for the river, but early winter is a hazardous time. The panicked animals might break through thin ice or become trapped in a water hole. Unable to extricate themselves, they give up the long struggle and succumb.

As winter progresses, the wise old cow leading the herd avoids the river except its outlet by Jasper Lake which is the only point where the water generally stays open. Elsewhere, if hard–pressed by wolves, the elk revert to a very different strategy. Prepared to sell their life dearly, they make a stand on a spot that affords protection for their vulnerable back side. Such shelter can be found against a dense clump of trees, on a cliff or the very edge of a precipice. Facing the enemy, they are ready to strike with their front hooves. Bulls use their antlers as a shield.

In back of the Lookout Hill is such a place of last resort. Here, on the upper lip of the canyon, the snow is often trampled by the footprints of elk. This puzzled me greatly at first, until it suddenly dawned upon me that this odd behavior and strange choice of place had to do with predation. Most times when I found massed elk tracks on the very rim of the cliff, there were also tracks of wolves nearby, either on the hill or on the meadows below.

One blustery afternoon, I saw a yearling bull at bay on a steep outcropping of rock. He was looking fixedly at the bushes framing the crest of the hill above. Following his gaze through binoculars, I discovered the head of a wolf, watching me. He was probably none too pleased with my untimely arrival. Presently, he sidled off into the trees. I recognized him as the dominant male of the local pack.

Backing away, I went around the base of the hill and walked to a high point from where I could see the young bull. Beyond, near the woods, lay the wolf pack, curled up and fast asleep. The wind was very cold and I soon decided to leave and go back to camp for a quick warm–up. Less than an hour later, the wolves and the elk had gone. I never found out how their interaction ended. The snow cover was intermittent and useless for tracking.

~

When the nights become frosty and the poplar leaves turn golden, elk venture out into the open, predators or no predators. It is rutting time and the bulls are in their prime. After a summer in solitary seclusion, eating and resting like prizefighters preparing for a great battle ahead, they bugle their shrill summons to the cows. To enhance their appeal, they urinate on their legs and on the ground, wallowing in it. Caked with mud and engulfed in his own kind of perfume, the randy bull repeats his seductive message. Driving his female admirers together in a tight harem, he jealously prevents them from straying and falling under the spell of rival suitors.

Mature bulls compete with each other and if they are a close match in size, it may come to blows. Clashing their antlers together and pushing for all their worth, they try to unbalance and gore their opponent. More often than not, actual violence is prevented by the visual impact of a superior, multi–tined rack, its ebony prongs gleaming like polished daggers. As a hallmark of virile rank, antler size alone can be enough to keep others of lesser stature at a respectable distance. The smaller bulls end up sneaking around the edges of the herd, biding their time to impregnate a straying cow after the dominant stag has lost some of his drive.

At the peak of his power, the overlord can take on any predator. If wolves attack, chasing the herd, they bypass and ignore the defiant bull as untouchable. Enraged, he may thrash a bush with violent blows of his antlers, but the cows have fled. To get his females back and keep them together, the bull has to be on his feet day and night, allowing himself no time to eat. In fact, in a month's time, his weight can drop by more than one hundred pounds (45 kg). After the rut is over and winter begins, he is severely weakened and at the mercy of his enemies. It is the paradox of the magnificent wapiti that the greatest of them all ends up the most vulnerable to wolves.

One bitterly cold December day, Peter and I came across a small meadow where the snow–covered ground looked like the arena of gladiators. There were tracks of wolves all over. Leading away from the spot was a trail of broken branches, here and there smeared with blood and littered with tufts of elk hair. Following the telltale sign, we came to the river which was partly frozen. Scoured free of snow by the north wind, the glare ice was dotted with a spoor of red drops that ended abruptly at an open channel. Evidently, the elk had entered the swift current, but where was he now? Looking around, Peter suddenly exclaimed: "There he is!"

Lower down, we discovered antler tines protruding out of the new ice. Under siege by the wolves, and unable to get back out of the water, the weakened bull had evidently succumbed to his wounds and the cold.

It was not until late March when the river began to open up and the carcass of the great wapiti partly thawed out of the ice. One day, I saw the wolves feeding on it, tearing and pulling up whatever they could before the leftovers were washed downstream by the waxing torrents of spring.

Over the years, I have found the skeletal remains of several other huge bulls, the spent harem masters, who had given their all during the rut and never had a chance to recover before the blizzards hit. I often revisit these sites. The bones are scattered by scavengers, but the heavy skulls lie where the bulls made their last stand. The great antlers form an ebony epitaph, an impressive reminder of the fragility of life. Death is also a renewal and life is eternal. Recycled by a host of creatures, from the wolf to the most minuscule worm and microbe, the bull's essence returns to the soil, feeding the flowers of summer. And his superior genes are preserved in the calves that he sired during the heady days of his greatest triumph, however brief.

CHAPTER 16

OF SNOW, MICE AND MOOSE

Like water, snow can be a blessing or a curse for humans and animals alike. As with other good things in life, the keyword is moderation. If the fluffy crystals accumulate to little more than ankle–deep, they do not impede pedestrian travel, but alternating cycles of thaw and frost make for noisy and tiresome going. Once the layer reaches to over the knees, every step becomes an exercise in frustration. In my greenhorn days, I thought that skis would allow for an easy and fanciful glide through the winter wonderland, but they are no magic wand. Unless there is a packed base, the narrow slats sink deep down, making the going even more laborious than without them. Breaking your own trail through the bush, perhaps following the erratic spoor of a fox or wolf, is very different from recreational cross–country touring on a groomed circuit.

As a novice skier, I once set out on an overnight trip, pulling a toboggan loaded with camping gear behind me through deep snow. I soon faced defeat and returned to my starting point much the wiser. Nordic skiing can even be dangerous as I found out the hard way when a firm crust allowed me to stay on top. Travelling on old–style skis fitted with rigid bear–claw bindings, I zoomed down an open, wind–packed slope until I reached a soft drift below.

My legs suddenly broke through, stopping me cold, head over heels. Had I fractured a bone, the accident could have had fatal repercussions. The temperature was far below freezing. There was a brisk wind, and no–one knew where this crazy Canuck had gone. Wiping the bloodied snow off my face, I henceforth approached any descent with the utmost of caution. Little did I know that there were still more rude surprises in store for me, even on the level.

One beautiful March day, a Banff Park warden gave me permission to drive part of the restricted road into the high country of the Cascades where I wanted to check for wolf sign. Snow lay six feet deep (1.8 m), but it had settled well and there was a firm crust on top. After parking the car, I strapped on the skis and took off cross–country. Avoiding steep terrain, I sailed along nicely across a slope, feeling on top of the world, until I lost my balance on a bumpy spot and keeled over sideways in a downslope direction. By this time, it was afternoon and the sun had softened the surface. Sticking out my arm to brace the fall, it speared right through the crust without touching bottom. Groping for support, I ended up almost upside down, my shoulders lower than the level of my feet. With the greatest of effort, I unbuckled my skis and finally managed to sit up and put them on again. All this while, two myopic moose, standing higher on the hill, had been watching me, trying to make sense of this sprawling octopus floundering in their sterile domain.

Much safer and more manoeuvrable than skis are snowshoes, either made in the time–tested way of wood and rawhide webbing, or of lightweight metal and plastic. Although you still sink well down into soft snow, a crust will hold you up. Ambling along at your own pace, you can stop anytime you want, and without the bother of poles you have your hands free to focus the binoculars if something of interest catches your eye. If you return the same way you came, you can make use of your own trail, which is easy as well as safe since you cannot get lost.

While the beauty and pleasure of snow are ours to enjoy, the intrinsic value of this wondrous medium is quite a different matter. In the northwoods, snow is a vital security blanket that insulates the ground, preventing the soil from freezing too deeply, which would otherwise delay spring even more. Its airy mantle is also a blessing for small, warm–blooded critters, which need all the help they can get to survive six months of winter. The bottom layer of the snow sublimates and withdraws, leaving a domed airspace near the ground, like a climate–controlled, science fiction city of the future. In this dead–calm netherworld, the rodents and weasels of the forest continue their feverish affairs protected from the worst of blizzards.

Deep snow is a safehaven for grouse as well. At nightfall, they dive headlong into its soft embrace and the birds tunnel to a cosy sleeping chamber protected from the chilling wind. To get about on the soft surface, grouse have evolved special footwear. Before the start of winter, their toes are broadened by stiff feathers growing outward. This design may well have inspired the aboriginal prototype of our snowshoes.

Another common boreal animal that Mother Nature has fitted with oversized feet to enhance the weight–to–surface ratio is the varying hare, commonly called snowshoe rabbit. By virtue of its huge clodhoppers, it stays practically on top of the softest powder. The deeper the snow, the higher the white bunny is lifted up, which brings more nutritious twigs and bark within reach of its buck teeth. As always, there is a catch. Their pathways give away their hiding places and lend support to stalking foes such as the fox and the lynx. These too are fitted with relatively large and hairy feet that spread their weight and help them get about in snow.

Like migrating birds, some boreal mammals dodge the problems of winter by retiring for the season. Bears and marmots hibernate underground. Beavers withdraw under the ice. The few species that remain active in the northwoods are forced to cope. Some do better than others. When the going gets tough, adaptations to snow and cold, either behavioural or physiological, can make the difference between life and death. A study in contrast are the elk and the moose.

~

Severe winters can lead to starvation for elk because, as grazers, they have to dig through the snow to get at the grass. The deeper the drifts, the more energy the animal needs to expend for every mouthful of fodder. By comparison, the menu of the moose is little affected by snow. Standing tall on long legs that are an asset in deep water as well as snow, a moose pulls down palatable branches from bushes and trees. Its long snout allows it to reach even higher. Scientists claim that its odd, pendulous nose is also an unique cold–weather adaptation. Its function is to pre–heat the frigid outside air before it gets to the lungs. Even the way a moose moves is a response to its habitat. The animal pumps its long stilts up–and–down like pistons so that its hooves meet with a minimum of resistance. By contrast, elk drag their feet, as shown by their tracks in the shallowest of snow.

Another significant difference between the two can be seen in the time the bulls retain their heavy headgear. Since antlers are mainly used as status symbols during the brief autumn rut, and since they are regrown the following summer, why continue to carry them around over winter? Again, the moose is the more sensible of the two. The bull sheds his palmated blades a matter of weeks after the October mating season. Elk, however, do not drop their thirty–pound (13.5 kg) rack until late March!

The two large herbivores differ also in social behavior and anti–predator strategies. In stark contrast to elk, moose forego the benefits of herd living. Instead, they isolate themselves from others in the thick of the woods and seek safety in seclusion. A loner for most of the year, the cantankerous giant is content to stay put in its restricted domain and it wanders as little as possible, so as not to draw undue attention to its whereabouts. For moose, the best defence against predators is to hide from them.

Yet, over vast expanses of North America, this well–adjusted denizen of the northwoods represents the major prey for wolves. How do they do it? First of all, to find the thinly–spread recluse, wolf packs need to travel far and wide. And secondly, they save energy whenever they can. If you follow their tracks, it soon becomes clear that the cursorial hunters choose a path of least resistance on frozen rivers, plowed roads, windblown ridges, and under the umbrella of evergreens. They tend to avoid the deep snow in open muskegs, which are the preferred retreat of the long–legged moose.

One overcast December day, I was tracking a pack of seven seeking their way in single file between the stunted trees of a tamarack bog. Plowing chest–deep through the snow, they stepped exactly in each others' footprints. After traversing the low ground, the wolves ascended a ridge where the snow was a little less heavy. The slope afforded a strategic advantage as well. If they sensed prey in the muskeg below, their attack would be assisted by gravity. At one point, the pack had come to a halt, huddling together. A little farther, they had sprang downhill in flying leaps and bounds. At the base of the slope, I found the circular bed of a moose. A wad of hair lying on the snow was proof that its attackers had been upon their prey just as it got to its feet. The chase led back across the muskeg, with the loping wolves weaving in and out of the deep furrow plowed by the moose. As evidence of their close contact, I came across a few more lumps of hair. The hunt carried on well over half a mile, but eventually the wolves had given up. Strung out along the route, I counted seven beds. The powerful moose had gone on like a bulldozer and gradually circled back to the area where its sleep had been so rudely interrupted. It had been lucky to make its escape and, by the looks of it, no blood had been spilled even though running away is a risky thing to do. Animals that flee render themselves vulnerable if the wolves follow closely behind. While some of them inflict deep bites to their victim's rump and legs, others try to slash the soft underbelly, spilling the guts. One wolf may even seize the moose by the nose. Trying to break its tenacious grip, the animal shakes its head and lifts the wolf bodily off the ground. The end comes after a long struggle, and the predators begin eating before their prey is dead. Such grim details have been reported by biologists observing kills from aircraft.

Moose that refuse to flee stand a much better chance of surviving wolf attack. Belligerent and ready to strike with their lethal hooves, they confront their enemies and save their hide, at least for now. Eventually, to quote old–time naturalist Ernest Seton–Thompson, the life of all wild animals inevitably comes to a tragic end. The older the moose, the more likely it is to become weakened by malnutrition or disease and to be overcome by its foes. The maximum age of moose is around eighteen years.

If the snow becomes hard–crusted as well as deep, even a healthy moose may get into trouble. Forced to restrict its radius, it stays in its own ruts, perhaps meeting up with others of its kind. Together, they create a mosaic of pathways called a moose yard. If left alone, they should be fine as long as the food supply lasts, but often it becomes depleted before spring releases the browsers from their white prison. The last weeks of winter are hardest. The crust can become strong enough to carry a wolf or a two–legged hunter.

In the past, snowbound moose were routinely tracked down by aboriginals on snowshoes. The mightiest of bulls that refuses to run, ready to ward off any wolf, stands no fighting chance against spear, arrow, or bullet.

~

Female moose that survive the hardships of winter soon face the ultimate test, when they have to defend their newborn progeny as well as themselves. One day, I saw a cow and her tiny calf run out of the woods to the open banks of the Snake Indian River. She took a few steps into the turbulent water, but her little one failed to follow. She hesitated and returned to shore. Presently, a black wolf emerged from the trees. He was alone and did not even bother to approach the cow. After nosing about for a while, he went back into the woods. The moose had stayed where she was, her calf standing under her belly.

If confronted by a group of wolves, a mother moose can put up a spirited defence. Naturalist Jill Seaton, who lives in Jasper, once observed a cow repel a pack for several hours. Standing in a pond, she kept her youngster at heel. Each time the wolves came too close, she rushed them, and her dapper little one followed right along behind mother. However, in other cases, if the predators succeed in pulling down her calf, the cow's courage can lead to her own demise as well since she is loath to leave the scene. It must be heart–rending to see such a drama in real life, no matter how much we try to remain objective. Moose often give birth to twins. Unable to defend both at once, she may lose one and run. However, the death of her only calf is a trauma a mother moose will not soon forget, or so it seemed to me when I found evidence of a recent kill. All that was left of the calf were a few shattered bones and a mat of hair on the snow. On top lay a fresh pile of pellets of an adult moose. Evidently, the cow had returned to the site, mourning her calf.

Where an adult moose has met its Waterloo, the evidence of a struggle can be quite dramatic. If the animal made its last stand in a pond, the shoreline vegetation is trashed. Once, I was shown a kill site near Devona where the carcasses of two adult male wolves lay nearby. Apparently, the cow had managed to stomp them to death before her time was up.

If wounded and pursued by wolves during winter, moose often flee toward the partly frozen river and become trapped in ice. One day, a cow that was bleeding from bites to her rump, had entered a belly–deep water hole in the Athabasca River. She succumbed during the long, cold night. By early morning, her carcass had been dragged onto the ice and was partly consumed. Snow flakes, floating down ever so gently out of the leaden sky, were covering up the gory evidence with an immaculate white shroud. In the dark woods, ravens were cawing a macabre requiem, their way of celebrating the continuity of life.

CHAPTER 17

RUN FOR "DEER" LIFE

On the meadows below the Lookout Hill, a herd of elk cows and their calves suddenly comes to attention, ears pricked up like antennae, all facing toward the woods. A moment later, they run into the opposite direction, bunching tightly together. But they do not go far and halt soon, looking back apprehensively. Their heads turn slowly as if they are following something that is running through the trees, hidden from view. Presently, they resume grazing as if nothing has happened. Disappointed not to have seen what I had hoped to see—a wolf or wolves streaking after the elk—I wait another twenty minutes. I then leave to go back to the cabin for breakfast.

Returning to the hill an hour or so later, I sit down on my makeshift seat and scan the flats and the frozen river. The elk have gone. The croaking of ravens calls my attention to the far corner of the meadows. Focusing the glasses, I pick up a couple of the black rascals over the trees. Is something going on there? Mindful of the elk's alarm behavior earlier this morning, I remember that the herd was looking into this very direction. A few minutes later another scavenger arrives, a bald eagle. As it perches on a spruce, its white head contrasts brightly with the evergreen background. The sharp–eyed bird, spying on corvid activity from afar, has come in for a share of the spoils.

My curiosity peaked, I go downhill, walk across the meadow and quietly enter the woods. Nothing moves, not even a raven. The eagle has already departed. But then I catch the telltale white–and–black flash of a magpie. Near the spot where it flushed, behind some bushes, I behold a sight that makes me catch my breath. The bloody rib cage of a fresh kill juts up over the grass, like a wet, crimson tent! It is a mule deer, the skin torn away. Only the innards and some of the muscle have been removed.

Who is the culprit? There are no tracks to give away the story, for it is late March and the snow is practically gone. Streaks of blood stain the still frozen ground. Examining the immediate surroundings, I come across a few small patches of crusty ice. One of them carries the bloody signature of the killer, a single paw print, red on white!

That afternoon and evening, I spend a lot of time on the hill. Winged scavengers fly off and on. At dusk, I catch a brief look of two wolves trotting across the meadow. Presently, I try a howl but there is no answer. Wolves seldom vocalize if they are on a kill.

Next morning, I check out the carcass. All edible matter is gone and the bones are scattered. There is no sign of the killers.

Although the entire chase and kill sequence had been screened from my view, I consider the above observation a lucky break. It represented one of the few times a successful hunt occurred in my vicinity, at least as far as I am aware. No doubt, others could have taken place while my back was turned or a matter of minutes before my arrival on the lookout.

One very windy morning, while Peter and I were walking in the lee of a hillside forest, we saw a black wolf dash off between the trees a dozen paces ahead. Focusing on its tracks, we walked right by its fresh kill lying just out of view a little lower down the slope. We found it on our return, half an hour later, alerted by a flushing raven. Investigating the spot, we discovered a dead white–tailed deer, still warm. One of its eyes had been gouged out, which is what ravens do first. Other than that, the carcass appeared to be untouched safe for a slight tear in its flank where the predator's fangs had torn the skin. The victim was a fawn, just over half a year old. During its flight, following its mother and upon emerging from the woods, it had been brutally intercepted. The cause of death might have been shock, arresting its racing heart, snuffing out its tender life like a candle in a gust of wind.

If you happen to flush a predator out of cover, it is always wise to go to the trouble of checking the area for kills, no matter how often your suspicion may prove to be groundless. One morning, just after he set out on a long hike, Brian saw a wolf run out of some bushes. Unwilling to dawdle, he walked on by. Next day, both of us went back there and found the remains of a white–tailed deer, all but completely consumed. Snow conditions were perfect for tracking, and since there were only two animals involved, one wolf and one deer, the story of the chase was easy to unravel. To make sure, we backtracked the wolf for about a mile. It had come down the Snake Indian Valley on an old park road which we call the Back Trail. At a point where the road leaves the forest and begins to descend to the flats, it crosses a sheep trail that runs roughly parallel to the road but stays higher up on the open slopes. The wolf had followed the sheep trail and suddenly bounded down to intercept a deer running below. Instead of entering a poplar grove where snow lay deep, the deer had skirted the edge of the road and was nailed by the wolf in an instant.

To find such clear and unequivocal evidence of what took place is in sharp contrast to the confusion of sign if a big pack of wolves chases a group of prey in wooded country. Not only is it all too easy to lose track of the hunters, particularly if they jump over and through bushes, the unknown factor is always relative position and timing. Which animal was ahead of the others? One wolf may have done all the chasing while others follow at a distance. Even if there are only two or three wolves involved, you seldom know for sure which one was the frontrunner. On old kills, the picture is complicated by "cold meat wolves" that shadow packs at a respectable distance, perhaps a day or so behind. In human terms, wolf–society is very class conscious. The lowest on the totem pole is not a predator but mainly a pathetic scavenger, satisfied to chew on the bones of the kill long after the top dogs have left.

During summer, lone wolves concentrate their hunting efforts on the smaller prey species, such as beaver, muskrats, ducks, and geese. However, after waterfowl migrate and aquatic rodents retire in their icebound lodges, the lone wolf's larder becomes lean. Rabbits or hares remain a year–round source of food, but they are seldom numerous in the predominantly coniferous forests of Jasper Park. In order to survive winter in the mountains, wolves need to hunt hoofed mammals. Since the larger ungulates such as elk and moose are quite capable of fending off a single attacker, lone wolves depend on deer that are not too hard to kill. The few people who have seen a capture claim that all it takes to slay a small deer is one bite to the head or neck. However, catching an alert and healthy white–tail or mulie is quite another matter.

Whereas the members of a pack can cooperate and take turns in chasing the prey, a lone hunter has to do the job alone. The odd capture may be made comparatively quickly, such as the hunt described above. Others require a long, exhaustive chase, the proverbial dogged pursuit. The longest known case on record covered a distance of at least thirteen miles (21 km). This hunt featured a white–tailed deer and a radio–collared wolf that was being watched from the air by a team of researchers in Minnesota. They witnessed the start of the chase, but not how it ended. After two hours of continuous observation, their aircraft had to return to base for refuelling.

Over the years, I have seen evidence of several long hunts and involuntarily interrupted at least two. In the first case, I was walking on an open ridge when the breaking of crusted snow in the woods below alerted me to a white–tailed buck that ran by between the trees. After he had disappeared from view, I looked back to where he had come from and discovered a lone wolf standing on a slight knoll. Evidently, he had spotted me and stopped. As it so happened, Peter was about half a mile (0.8 km) farther down the ridge and the buck passed him right by, continuing his flight. The animal looked exhausted and disoriented. Instead of going around some bushes, it struggled through. Descending the ridge, the buck slipped and fell on a patch of ice, where he lay prone for a few moments, licking his bleeding front leg. The skin must have been cut by the hard surface of the snow. After the buck got up, he resumed running and went out of sight. Little did he know that his terrible pursuer had already been stopped.

A week or so later, on the same ridge, a mule deer doe came bounding right toward me. I stopped and she approached closely, halting a few feet away. Her mouth was agape, tongue protruding. Her eyes were glazed over with fatigue and fear. After a moment of hesitation, she darted by at arm's length and resumed her flight.

Suspecting that she was being chased, I stepped aside behind a spruce and waited. Seconds later, a grey wolf, red tongue lolling, came loping down the trail. A dozen feet off, he became suddenly aware of my presence and spun around, dashing back the same way he had come.

Thus, twice in as many weeks, I had saved the life of a deer and left a wolf hungry. I was not sure whether to feel good or bad. On the one hand, I hate to interfere with the lives and destiny of wild animals. On the other, I was mighty happy to have witnessed such a telling example of prey and predator running for dear life.

Chapter 18

The Foxy Underdog

The fox came down the path in the light–footed trot typical of his tribe, like thistledown blown on the wind. Abruptly, he turned aside and stood at attention, ears cocked. Burnished by the low light of the winter sun, he looked resplendent, his rich fur a blaze of orange, offset by the velvety black of ears and legs.

Suddenly, like an uncoiling spring, he jumped up high on all fours and sailed through the air in an elegant arc. He pounced down hard into the snow, front feet first. Next instant, he raised his head in triumph, a mouse clamped in his jaws. Prancing about, he selected a spot where he deftly dug a small hole. Then, in a quick routine, he deposited his prize and covered it over with snow, using his nose instead of his feet.

Before he had gone on more than a dozen yards, he performed another high–flying pounce. His front feet hit the ground like a spear and he again pinned down a tiny prey. This one was also cached at once. Returning to the path and heading my way, he broke into a happy lope, until he suddenly spun around in a panic. He cleared the tall grass of the ditch with one bound and was gone between the trees. The last I saw of him was the white tip of his long brush, floating out of sight like a butterfly.

Tracks led me to both caches. The first contained a fat vole, the second a tiny shrew. The latter is not a rodent but a predator in its own right, the smallest carnivore on earth. It hunts insects, worms, and even baby mice. Its shrill squeaks and rustling in the ground litter present the fox with an irresistible target. Yet, a shrew is seldom eaten because of its skunky taste. The fact that this one had been buried was therefore no great surprise. However, the fox had also cached the vole, his staple food. This indicated that he could not have been hungry at the time he had made these kills.

~

Like most predators, foxes lead a life of feast or famine. If the opportunity presents itself, they kill in excess of their immediate needs and store the surplus for a rainy day. Later, passing by the same trail, they inspect their string of caches like a miser gloating over treasured possessions, consuming them only if needed.

By their proven ability of finding their caches back, foxes demonstrate that they have a good memory. They even employ a simple bookkeeping system that tells them at once if a cache has been emptied. This fascinating discovery was made by researcher David Henry during his studies in Prince Albert National Park. Henry noted that a fox, after eating a stored item, often voided on the spot. Next time the animal made his rounds, the odour signalled which caches were empty, even though they might still exude a residual smell of the food they had contained. The fox was spared the effort of digging for nothing by the presence of his own scat or urine.

Apart from its use in storekeeping, urine is also a ready medium for territorial messages. Staking out the boundary of their domain like a fence in suburbia, resident males frequently aim a quick squirt at prominent sign posts such as stones or plant stalks. Piddling every few dozen yards or so, they seem to be able to measure out their supply with precision!

I was familiar with this compulsive scent marking habit of the red fox from the Alberta farmlands, where they are locally common. But in Jasper they were considered rare and seldom reported. It took me years before I found definite evidence that the species occurred at all.

As with all mammals that are largely nocturnal, the best way to check for foxes was to look for tracks, but this was not as simple as it may sound since their spoor closely resembles that of the coyote.

The latter is very common in Jasper Park. Three times as heavy as the fox, it is Reynard's major enemy and competitor. Locally, a saturated population of aggressive coyotes leaves little room for foxes, particularly in the lower Athabasca Valley. Therefore, the first time I found definite proof of their presence stands out as a red–letter day.

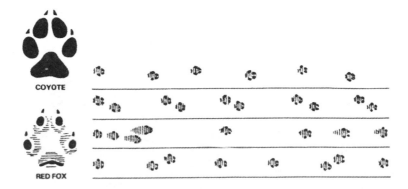

Track patterns of the red fox. Fom top down: walk, trot, lope and "fox trot". Note the small toe and heel pads of the fox compared to the coyote.

On this happy occasion, Peter and I were snowshoeing up the Back Trail behind the Snake Indian canyon. A small canid had gone ahead of us for some distance and every once in a while it had meandered to the side of the path where it had left its mark on tufts of grass or conifer seedlings. Could this be a fox, I wondered? If only we humans had better noses! Foxes are supposed to be quite smelly. Some people claim that they can detect their scent on a fresh trail. But first of all, you have to know what a fox smells like. To find out, I bent down and put my nose close to a tiny spruce that had received a generous dose of yellow droplets. As human olfactory powers go, mine is poor, perhaps ruined by too many years of pipe smoking. Yet, a musky stench was quite noticeable. While I repeated my test at the next couple of urine stations, Peter looked on without comment, but somehow the word spread. Many years later, I was introduced to a Jasper resident who reacted as follows: "Oh, you must be the guy who smells fox pee!"

Walking on along the path, I began to scrutinize each and every paw print, examining them for the one foolproof clue that serves to identify the fox. As stated before, its track is quite similar to a coyote's although the latter is generally a touch bigger. Yet, a big fox may leave a larger print than a small coyote. Size alone is therefore of little help. The main difference between the two is that most of the foot sole of the fox is covered with hair, leaving only tiny bare pads, whereas the coyote's foot has large pads, like any dog. Since canids lose heat through their soles, the furry feet of the fox serve as an adaptation to the cold climate.

Continuing my search, I found most of the prints too fuzzy for identification and little more than a hole in the snow. But on an open section of the trail, where the surface was wind packed, the animal's foot had registered like an exquisite engraving, clearly featuring the small toes. The heel pad showed like a narrow transverse ridge, which is one hundred percent diagnostic for the species. No doubt about it, this was my first mountain fox!

There were additional, more subtle, clues. A nine–pound (4 kg) fox can walk on top of drifts in which a thirty pound (13.5 kg) coyote flounders. The fox's relatively large foot size is thus another adaptation to boreal winter conditions. His flexible toes spread apart to further increase the width of the paws. Sneaking along in a slow dancing motion, feeling their way, foxes find support on the slightest crust. Catlike, they seem to enjoy walking on fallen tree trunks. Rambling along, they switch their gait often, like a school child on recess, springing and dancing. The happy–go–lucky fox is a ballerina in a fur coat.

~

Ever since that first find, which alerted me to the local occurrence of my foxy friends, I began to pay more attention to the tracks of small canids. Over several decades an intriguing pattern has emerged. Nearly all fox sign was located on the semi–wooded hills or along their base, and very seldom on the flats. The foxes appeared to have an affinity for steep ground. Their tracks went up or down snowy river banks that seemed all but inaccessible for any mammal. One dusky winter evening, looking down from the rim of the canyon, Brian and I spooked a fox on the ice below. To our great surprise, he dashed right up the nearly vertical sandstone cliffs on the opposite shore! Its significance dawned upon me when I realized that this fox would probably have reacted in the same manner had he been chased by a coyote. On steep terrain and on deep drifts, the nimble fox, by virtue of his light weight and big feet, has an excellent chance to get away from his heavier cousin.

Strife between the two was well–known to me from my observations on the Alberta farmlands. There, coyotes are ubiquitous on the more secluded fields and pastures. This forces the underdog to make do with the leftover habitat near houses and roads. If confronted with their wild cousin, or with farm dogs for that matter, adult foxes can take care of themselves. Blessed with superior speed, they outrun their larger enemies and dodge astutely. But the pups are vulnerable. One afternoon, while I was watching six youngsters at play, I heard the hoarse, peacock–like scream of the vixen. In the nick of time, her puppies dashed back underground just ahead of a coyote that came racing out of the nearby bushes. A few days later however, the den was deserted. As silent evidence of the drama that had taken place, I found the tail of one of the pups nearby.

To get away from their nemesis, the clever prairie foxes have learned to establish their dens on the edge of farm yards or busy roads where coyotes are less common. Thus, by and large, the two are spatially segregated. A similar arrangement proved to exist in Jasper. The mountain foxes avoid their adversary as much as possible. While the coyote dominates the grassy flats and the riparian bottom lands in the Athabasca Valley, which are richest in small rodents, most foxes stick to the hills. Here, if pressed hard, they can find refuge on steep terrain.

My hypothesis was supported by a comparison of the respective location of sightings of the two species. Of the two dozen foxes I have seen at Devona, all but two were on or near the hills. Conversely, nearly all of 120 coyote sightings were centred on the flats. This segregation of habitat was also evident in the vast majority of tracks I found.

Over the years, it has become plain that the fox population in the Devona region is subject to severe ups and downs. This is to be expected, since this predator is dependent on adequate stocks of prey that are cyclic in occurrence. In the predominantly coniferous forests of the Athabasca Valley, mice and voles vary in abundance. So do snowshoe hares, squirrels and grouse, which are thinly–spaced at the best of times. During summer, the fox's menu can be supplemented with grasshoppers and raspberries, but by fall his choice becomes limited. As indicated by the content of scats, I found that the local foxes subsisted on the dry fruits of kinnikinnick or bearberry. How much nutrition this could provide is open to question, for most of the stony berries appear to be swallowed whole and pass through the fox's alimentary tract undigested.

When hungry, foxes readily take advantage of carrion, which can get them into all sorts of trouble. Tempted by the ripe odours of a wolf kill drifting up from the flats, they may throw caution to the wind and venture out from their hillside refuge.

One evening, I heard the hoarse alarm of a fox on the meadows. Presently, looking over his shoulder, he came running to the hills. Expecting to see a coyote as the cause of his fear, I scanned the flats and spotted a wolf instead. A few days later, on the shore of the Athabasca River and not far away from the remains of a moose calf, I found a dead fox, his slender body mangled but not eaten. The surrounding area was crisscrossed with tracks of his relatives, wolves as well as coyotes. But I will never know which of the two had killed the pretty, little underdog.

Chapter 19

Cousin Coyote

From the Lookout Hill I had a clear view of the messy remains of an elk bull lying on the frozen river. He had died in early December when the local wolf pack chased him into the last open water hole. Wounded and unable to get out, the animal had succumbed. During winter, the wolves had occasionally inspected their property and sprayed their yellow signatures against the antler sticking up out of the snow. Now, in late March, the ice around the cadaver was melting. Absorbing the sun's heat, the frozen river would soon break up and flush the bull's remains into a watery grave.

In the meantime, the carrion provided a smelly magnet for hungry scavengers although the pickings were lean. All edible matter within reach had already been removed by ravens and wolves. Earlier in the morning, I had seen a pack of eleven tugging away at the rib cage and scattering the turf–like stomach contents. Now, during their absence, a coyote had sneaked in. He was partly obscured from view in the cavity that had formed around the thawing carcass.

Suddenly, three black wolves sprinted out of the nearby woods. Fleeing in the nick of time, the coyote streaked away across the ice. The larger of his pursuers, bounding in giant leaps, was no more than a body length behind when the coyote vanished from view between shoreline trees. Presently, two of the wolves came back and lay down on the ice. The bigger wolf stayed away another ten minutes or so, but I never found out whether or not he had been successful in punishing the interloper. Snow conditions were too poor for tracking.

Coyotes commonly pay the ultimate price for interfering with wolf kills or just for being in the wrong place at the wrong time. Numerous mortalities have been reported from Jasper Park and elsewhere. A big western wolf is four times as heavy as the average coyote. Little wonder hostilities can be fatal for the latter. However, in eastern Canada and the northeastern States, where the smallest wolves weigh little more than a big coyote, the two species have been known to interbreed. Genetically speaking, *Canis lupus* and *Canis latrans* are close cousins and their intertwined origins are an ongoing subject for debate between zoologists. A mating between them can produce fertile offspring. Either one can also hybridize with domestic dogs, but never with foxes. The chromosome number of *Vulpes vulpes*, the fox, is different from other members of the canid tribe.

In the past, before Europeans settled this continent, the running battle between wolves and coyotes was not as widespread as it is today. Formerly, their common home ground overlapped only in the southwest. The coyote was definitely not a boreal animal. However, in the last century, after mankind had decimated the ranks of the top dog, the wily coyote saw its chance to expand its range far to the north and east.

Of medium–size, this versatile and adaptable carnivore can subsist on a wide range of prey, from mice to deer. Thus, it competes for food with both the wolf and the fox. In regions where wolves are absent, their niche is filled by coyotes that specialize in taking down hoofed mammals. Although their weak jaws are not quite up to the task of making a quick kill, certain individuals become deadly hunters. Like wolves, some coyote families operate in packs of half a dozen that chase and harass deer persistently, particularly in winter. At Devona, bands of bighorn ewes with lambs race for steep ground as soon as their treacherous enemy approaches.

One July day, I stopped the car on the Jasper Park highway to take a photograph of a lone coyote trotting along the shoulder. While I was focusing, the animal suddenly froze, his lithe and scrawny body rigid like a statue. Following his fixed stare, I discovered a lamb on top of a steep, open hillside some two hundred yards ahead. As I found out later, the little one's mother was grazing with other ewes on the opposite side of the road, out of my view, but the lamb, from its high vantage point, could certainly see them. Left alone and isolated, it was calling repeatedly, apparently afraid of the highway traffic.

Riveted by the menace of the scene, I watched and waited. After ten minutes or so, the coyote began to stalk with slow deliberate steps, gazing straight ahead and ignoring several onlookers who had stopped their cars and emerged with camera in hand. Unaware of its peril and bleating pitifully, the lamb descended a ways toward a point where the slope was less steep. Below, hidden in a patch of tall clover plants, the coyote was lying up in ambush. After another ten minutes, I walked a little closer until I could just see him crouched down. Perhaps triggered by my approach, he suddenly rushed out of cover. The startled lamb fled downslope but was quickly intercepted and grabbed in its flank. Turning back uphill, it briefly broke free, to be seized once more. Held fast by the coyote, the lamb reversed direction again and ran down until it fell on the pavement of a picnic site at the base of the hill. A young woman, watching in horror, shouted and threw a stick, whereupon the coyote let go of his struggling victim. Fleeing one last time, the lamb did not get far. Bowled over, it slid down until it lay still and the predator shifted his grip to the throat.

After the prey's spasms stopped, the coyote held on for another five minutes or so. He then attempted to lift the limp body and drag it away. Unable to do so and possibly disturbed by the proximity of several people, the coyote abandoned his kill and trotted off to the highway. He briefly hesitated on account of a passing vehicle, then crossed over to the wide, grassy shoulder on the other side. There, he abruptly stopped dead in his tracks, looking fixedly at several sheep grazing near the edge of dense bushes some forty yards (35 m) away. The group included four ewes and one lamb. All of them seemed unaware of their peril until the coyote charged. The startled band split up and fled down the pavement, all except the lamb. Hotly pursued, it narrowly dodged the coyote's initial assault and entered the bushes, one jump ahead of its resolute pursuer.

Although I briefly searched the swampy woods, I could locate neither the killer nor his doomed victim. Not wanting to interfere further, I soon retreated, subdued by the violent yet natural drama I had just witnessed.

~

But exactly how natural was it? Undeniably, the animal actors in this tragedy belonged here in these mountains, and their struggle for life was ageless. Predators were killing prey long before mankind had the privilege of staying on the sidelines. Yet, in this particular incident human activity had created a multi–toothed trap. First of all, the fact that the lamb had been left behind by the band and had become vulnerable was directly related to the busy traffic. Secondly, once the coyote had made his first kill, the close presence of people was probably the reason why he had abandoned the carcass and attacked the other lamb.

Thirdly, the presence of sheep along the highway at this time of year—mid summer—was not quite normal. Traditionally, these ewes should have vacated their winter range in the lower valley two months ago, and retreated to alpine heights. There, they were supposed to withdraw into the roughest terrain before giving birth to their lambs. Instead of abiding by this age–old routine, this small band had remained near the highway. Why?

Again the answer must be sought in their deep–seated urge for security. This too was an anti–predator behavior, acquired in very recent times as a consequence of increasing highway traffic. Sheep feel quite safe near the road since the presence of humans keeps most wolves, grizzlies, and cougars at a distance. Moreover, sheep suffer from a morbid addiction to the salty chemicals used in snow removal. Eagerly licking the substance from the pavement, they tend to ignore everything around them, including people, cars and trucks... even their own lambs!

By the same token, the roadside represents a refuge for the coyote as well. He too appreciates a safehaven from his most dangerous enemy, the wolf! Over the decades, I have often been struck by the negative correlation between the occurrence of these two canid cousins. While the wolf rules the roost in the backcountry, most coyotes stick close to the highway corridor in the Athabasca Valley. Quite apart from safety, there are other substantial gains to be had: the chance for scavenging on traffic kills, and for taking confused sheep by surprise! Since the fumes and smells of the highway mask the natural world around them, the sheep become vulnerable to danger. And the wily coyote, the supreme opportunist, presses his advantage!

~

Another noteworthy aspect of the above incident was the lack of maternal care displayed by the sheep. Why had the ewes not defended their youngsters? Although the first kill might have happened out of their sight, the second attack was directed at the entire group. The ewes fled, not to be seen again, abandoning the little one to its fate. Although this may seem callous and unnatural to us, such a lack of collective defence against predators is quite typical of bighorn mothers. As mentioned in a previous chapter, it makes sense from the point of view of species' survival. Ewes that fail to run and instead turn back to protect their youngsters, run the risk of being killed as well. Evolving in a very harsh environment, sheep have adopted a communal society where the welfare of the group is paramount and ultimately benefits the individual.

In contrast to sheep, the mothers of solitary ungulates such as deer and moose behave in the opposite manner. They fight fiercely to defend their progeny which represent the future of their individual genes. One day, I watched a confrontation between a coyote and a white–tailed doe. Standing guard on a small river island, she rushed forward each time the coyote tried to sneak by her. He scurried back, tail between his legs. No doubt, there was a tiny fawn hidden in the dense, riparian vegetation. The dapper deer's resolve in her desperate fight was heartwarming to see. Next time, however, she might not be so lucky as to face just a lone adversary….

Coyote predation on the young of hoofed mammals becomes all the more deadly if they hunt in packs. In Yellowstone National Park, groups of three or more gang up on elk that have just given birth. Dwarfing her enemies, the cow is quite capable of stomping the devil dogs into the ground, but she is swarmed by their number and frustrated by their agility.

Some coyotes are quite capable of killing mature ungulates. At Devona, I came across the fresh carcass of a female deer that had apparently been dispatched by a lone coyote. Watching him on his kill, I could not help feeling sorry for his victim. A few weeks later, in the same area, Brian saw a coyote on a dead ewe. This coyote, probably the same individual in both instances, was obviously a determined hunter, showing what this medium–sized predator is capable of, given half a chance.

In eastern provinces and states, where "timber wolves" are absent, the so–called "brush wolf" has become the major killer of deer. Reportedly, the prey is routinely gripped by the throat and choked to death, apparently quite like the lamb I saw killed. For me, that particular incident represented the first time I ever had a close view of a predator killing a prey larger than itself, and I was horrified by its savagery. What if such ferocity would ever be turned against us? This is not just idle speculation, especially in regions where coyotes are tolerated and have lost all fear of humans.

Coyote attacks on people differ from vicious dog attacks in two major respects. First of all by their exceptional rarity compared to the countless maulings committed by our domestic pets. In North America alone, in an average year, dogs bite more than five million people, a quarter of them children. Several dozen die. Others end up with horribly disfigured faces! Coyote aggression, however rare, may be even more terrifying than dog attacks, because this renegade predator considers human beings as potential food!

A telling incident occurred in the Jasper townsite in April of 1985. Just after a mother had sent her two–year–old girl into the backyard of her house to play with a friend, she happened to look out of the kitchen window and saw the child seized by a coyote that had entered the yard by crawling under the rail fence. By the time the mother got outside, the coyote, holding its victim by the neck, was dragging the limp body into adjacent bushes. At the approach of the screaming woman, the animal dropped the unconscious baby girl, who was rushed to hospital and fortunately recovered.

In August of that same year, a coyote attacked a four–year-old girl playing in a Jasper campground. Alerted by her screams, the parents interfered just in time, before the child was pulled out of sight in the woods. The girl received severe lacerations to her face.

In July of 1988, there were two incidents in British Columbia in which coyotes attacked small children walking and playing in a provincial park. In the most serious of these, the victim was an eighteen–month–old toddler. When her dad rushed up, the child was face down on the ground and the coyote was biting her in the neck, causing massive bleeding and injury. In the other case, the coyote sneaked up on a four–year old, slashing her in the legs and buttocks while she ran screaming. The animal bared its fangs at other children who tried to interfere. Eventually, it slunk back into the bushes. In the same year, over a period of several days, a coyote bit and wounded five young adults sleeping outside their tents on a campground in Banff National Park.

Attacks on people, including adults as well as children, have been reported with increasing frequency from other localities, such as Yellowstone National Park, Los Angeles, Toronto and Vancouver. No doubt, there is more trouble to come. In fact, the phenomenon may be just in its infancy considering that it is only recently that *Canis latrans*, this most adaptable member of the canid tribe, has crowned its conquest of the continent by invading our cities.

CHAPTER 20

SECRET GO THE CATS

As I walked quietly through the woods, a cougar suddenly sprang out of hiding and long–tailed it out of sight. My first reaction was a jolt of fear. The secretive cat must have been aware of my approach and crouched in its lair until the last moment. Luckily, it had recognized me as out of bounds to its kind. Cougars are known to have attacked humans, whether out of ignorance or as potential prey. In the past decade, in Canada and the western United States, there have been three or four serious maulings each year and at least ten children and adults have been killed. The most recent Canadian fatality, in January 2001, was a female cross–country skier who was jumped from behind along a trail in Banff National Park.

In the mountain valleys of Jasper Park, this furtive feline called cougar, panther, puma or mountain lion, is not uncommon and may skulk anywhere. Undoubtedly, they have seen me more often than I have seen them. For every one of these big cats that I happened to flush, several others may have let me walk by, blissfully unaware of being within a few pounces of their mercurial power.

The most rewarding observations were those in which I happened to spot the cat before it saw me. One day, Peter and I were hiking down a sheep trail paralleling the Snake Indian canyon when we heard a repeated, bird–like call coming from the treed slope above us. Thinking that it might be an owl, we were much surprised to spot a cougar instead. Emerging into the open on the crest of the rimrock, it seemed oblivious of us. Presently, it went out of sight beyond a fold of terrain. Proceeding in stop–and–go fashion, we kept an eye on the far end of the ridge, examining it through binoculars.

"There he is!"

"Where, where, where?" I asked eagerly.

"You can just see his head between the stones! He is looking at us!"

On the crest of the slope, just above the place where the sheep trail would have taken us over into the next draw, the cougar was waiting! It struck me how his round head with the stubby ears resembled the boulders! This was near–perfect mimicry! Keeping us in his steady gaze, the animal clearly intended to waylay us. Yet, perhaps foolishly, we were not worried, expecting him to slink out of sight well before we got near.

Keen on a more detailed look, I suggested: "What if I go around the back of the ridge? Give me a few minutes before you walk on and scare him off."

I descended a ways down the north side of the hillside, which was wooded and deep in snow. A few minutes later, along the edge of a lightly–treed draw, the lithe lion, long tail swinging, came running downhill and streaked by in all his tawny beauty, brightly illuminated by beams of winter sun. Unaware of my silent presence, he vanished into the woods below. If only I could have captured that fleeting moment on film! But I seldom bother to carry a camera.

Anxious to report to Peter what I had seen, I worked my way back up to the top of the summit. Just as I emerged from the trees, he suddenly pointed behind me: "Look out! There he is!"

Turning round, I was startled to see the cougar a few yards below me. In the next instant, he dropped out of sight in the trees. Obviously, he had trailed and overtaken me as I struggled up the steep slope. How vulnerable I had been, at times hip–deep in the drifts, and how lucky I was to be human, inhibiting his fury....

~

Today, these elusive predators are commonly raised in captivity and made available as animal actors to photographers and film crews. This explains why action footage of cougars is commonplace on our television screens, but who has ever seen a wild one capture its prey? I have not. But it could happen anytime, or so I keep hoping. While watching sheep on the open slopes above the canyon, I have often fantasized about charging lions. It's a matter of being in the right place at the right time. One day, warden Brian Wallace stood on the same viewpoint and counted the sheep that were grazing or resting peacefully on the hillside. Averting his eyes and looking back again a few minutes later, he saw that all of them had sought refuge in the cliffs above the river. And near the top of the slope, Brian discovered a cougar dragging its fresh kill into cover!

One day, I came close to a similar event. Just as I arrived within view of the hill, a band of sheep raced up to the rimrock. Suspecting that the cause of their panic was a predator, I scanned the slopes through the glasses and vainly waited for ten minutes or so. I then cautiously approached the base of the hill and a big tom suddenly bolted from a cleft in the rocks. Perhaps he had already hidden his catch there, but I was unable to reach and investigate the spot.

Another day, a cougar had killed a bighorn below the crumbling cliffs of the Ram Pasture, which had offered inadequate refuge from this agile predator. When I saw him walk away from the carcass, he was accompanied by a black wolf, which probably was after a share of the spoils. Ascending the hillside a dozen paces apart, the two looked like a very odd couple indeed! Suddenly, the low–slung cat dashed at his long–legged competitor, which dodged no farther than necessary. During the following night, the cougar dragged the entire carcass up the hill and cached it in the woods, covering it over with some debris. Following his tracks, I again had the briefest thrill of seeing the cougar bound away through the brush. What would have happened had I been a wolf?

The enmity between these hostile clans, the canids and the felids, is a reality of their everyday lives about which we know little. So much of what goes on in the woods and in the dark of night remains hidden from our eyes! Based on a few incidental reports from field biologists, a cats–and–dogs altercation can seesaw either way. Cougars have been known to kill radio–tagged wolves, perhaps after they interfered with the cat's prey. Montana wolf trackers twice found evidence of the reverse. One of the victims was a small yearling cougar, the other a female. At a maximum recorded weight of 270 pounds (120 kg), a mature tom should be invincible, weighing twice as much as the heaviest of wolves.

A peculiar weakness of cougars is that they are easily tracked and cornered by hunters with hounds. Seeking an easy escape from his barking pursuers, the snarling cat climbs a tree and is shot after the people arrive. Yet, cougars routinely kill domestic dogs as well as the odd wolf or coyote. They have also been known to destroy wolverines and otters, apparently not to eat them, but just out of malice. By the same token, this cantankerous killer probably shows no mercy for his own kin, the other wild cats of North America.

~

The bobcat, an inhabitant of the arid south, is absent from Jasper, but its close cousin, the lynx, is a typical nordic aboriginal that is widely distributed. Although many lifetime hikers have yet to see their first one, I have found the lynx less secretive than the cougar for it often travels in the daytime. Moreover, engrossed in its own affairs, this cool cat tends to ignore humans if all you do is watch.

On a foggy November day when the bushes were thickly frosted, the yapping of several coyotes caused me to halt on the edge of a meadow. Thinking that the object of their alarm was a bear or wolf, I was much surprised to discover the ghostly figure of a big lynx. Unhurried and unconcerned about the obnoxious trouble makers, he chose a direction toward my vantage point. Crouching down onto one knee, I watched him approach, pacing along steadily. When I pursed my lips and sounded a mousy squeak, he changed course and came right up to me, pausing a dozen steps away and fixing me in a sphinx–like gaze.

This was another of those magic moments when I should have had a camera! But perhaps better not. The smallest movement on my part might have broken the spell. Instead, after considering me gravely for a few seconds, this haughty creature dismissed me as if I were of no more concern than a stump or stone. Turning away, he strode off into the hoar–frosted woods like a figure from a fairy tale.

Apart from the aloof, enigmatic expression on his whiskered face and his jaunty ear tufts, the cat's most striking feature were his huge, furry feet. They looked like dustmops. It was hard to imagine that these hairy slippers contained a set of deadly, retractable weapons.

The lynx lives by his feet. They carry this inveterate traveller over drifts where other predators sink deep. And the currency by which this boreal cat measures his material wellbeing, is the snowshoe hare. Like ours, theirs is a ruthless economy of supply and demand, but unlike ours, nature's cycle of ups and downs is everlasting and predictable. It is also without pity. Every ten years or so, the web of warm–blooded boreal life is severely shaken. At the bottom of the cycle, there may be no more than a dozen hares left where hundreds, even thousands, were scurrying about a few seasons earlier. The primary cause of the oscillations is still a matter of disagreement among zoologists. Some claim that the hare population crashes after the bunnies have exhausted their food supplies. Moreover, in mute defence to repeated browsing, poplar bark is said to develop a toxin that repels hares, exacerbating their plight. In contrast, other long–term researchers, such as Lloyd Keith, think that the cycle is driven by predation. As the hares increase in number, so do their many enemies, not only the four–footed hunters of the forest, but also hawks and owls. Raising large broods, their combined pressure eventually overwhelms the food resource. As the hares decline, the predators switch to other prey, such as grouse and squirrels, which subsequently decline as well. By early winter, when the tide of animal life reaches its lowest ebb, the doomed hunters turn on each other, precipitating their demise, while the last of the hares cower in their refuges. Delivered from tooth and claw, the promiscuous bunnies are destined to recover quickly, producing two to four litters of leverets next year.

Several recent field studies have divulged some startling facts about the harshness of life in the Canadian northwoods. In the world of the lynx, fat years are inexorably followed by hungry, desperate times. In the Yukon and Northwest Territories, following the crash in hare numbers, lynx abundance drops from thirty to less than three per one hundred square miles (250 km^2). Many succumb to starvation, others desert their range. Lynxes fitted with radio–collars have migrated south for distances of up to 660 miles (1100 km).

When stressed for food, cats do surprising things, such as killing foxes, even the odd coyote, for food. Despite their relatively light body weight of maximally thirty pounds (14 kg), they may take on much larger prey such as caribou or mountain sheep. Pouncing on the ungulate's back from behind, they cling to its neck in a lethal embrace, biting the spine and riding their victim until it stumbles and falls. Some lynxes, that were radio–tagged in Kluane National Park in Yukon, have been known to attack and cannibalize their own kind!

Caught in a vicious circle of violence, any lynx that is in poor physical condition or fails to secure a quick escape into a tree, may be done in by its canid nemesis, wolf or coyote. When winter begins in earnest in the unforgiving northern wilderness, the line that separates prey from predator becomes blurred like the hoarfrost that collects on the finest of branches after a foggy November night.

Chapter 21

The Elusive Otter

The expression "absence makes the heart grow fonder" partly explains the high value I place on certain species of animal that are now common in the Canadian Rocky Mountains but four decades ago very hard to find. Foremost in this select group is the river otter. Like all fur–bearers it had been practically trapped out earlier in the century, and during the mid 1950s the remnant population in western Alberta received a glancing blow from the extensive poisoning campaigns of that period intended to destroy wolves.

It was therefore a red–letter day when I, for the very first time, stumbled on the tracks of otters in Jasper Park. The year was 1966 and the location a muddy creek bed between Talbot Lake and the Athabasca River, by the outlet of Jasper Lake. The sign was clear and unmistakable, exactly like the drawings in Olaus Murie's *Field Guide to Animal Tracks*, which is a treasure house of information. The five–toed foot prints were grouped together in jump patterns typical of this low–slung water weasel. There were several sets of spoors, coming and going, some larger than others, indicating that this was a regular travel route of a pair, or perhaps even a family. In my mind's eye, I could see them loping along and my spirit soared at the prospect of setting eyes on this elusive creature in the near future.

During a subsequent visit to the mountains, Irma and I launched our canoe in Talbot Lake to explore its wooded shores and marshy bays. On the muddy bank of an island, pockmarked with otter tracks, we discovered that the lush grass had been flattened and was littered with scat. Otter excrement is one of a kind. The British have a special name for it: spraint. It is a loose package of course matter, blackish if fresh and bleached white if old. Poking it apart with a stick, we noted that it contained a variety of undigested matter: the scales of fish, bits of bone, fragments of beetles and mussel shells, and the occasional duck feather. All in all, it gave a rough indication of the diet of this aquatic predator. However, the most interesting thing was the scat's unique smell, fishy for sure, but not at all unpleasant. To this day, I rather enjoy collecting a piece of otter spraint on the end of a stick and holding it close to my nose for a whiff of its delicate perfume!

Exploring further, Irma and I noted that the otters had access to an intricate labyrinth of hiding places along the island's shore. Several holes and caves under the exposed roots of trees looked like possible dens. The British also have a specific name for an otter den: a holt. Anyone of these island holts might have been occupied since the grassy bank nearby was obviously a well–used playground and resting spot.

The afternoon was getting on. It had begun to rain and we needed to find a suitable campsite, for it was our intention to stay here for two days. On the east side of the island, out of sight of the highway that skirts the western edge of the lake, we located a protected spot for the tent. Soon we had a campfire going, its smoke disintegrating in the wind–driven rain that was getting worse as the evening progressed. We retired early and intended to begin our watch at the holt the following morning. Unfortunately, we had made one important mistake; we had failed to notify park authorities of our plans to stay overnight.

As darkness fell, a patrolling warden investigated our parked car which seemed abandoned. He deducted that we had launched a canoe into the windswept lake during the day and failed to return. His conclusion was swift: we had drowned! Alarm bells went off at park headquarters, but it was too late in the evening to start looking for us now. The search was postponed until next morning. Little did we know….

At the crack of dawn, over the patter of rain on canvas, we heard the throb of an engine. Sticking my head out of the tent,

I saw a small boat approaching, carrying two wardens. Dressed in yellow ponchos, they sat hunkered down under their wide–brimmed hats.

"Hi there! Good morning!"

They appeared to be much relieved to have found us alive and well instead of floating belly–up in the shallows! We, on the other hand, were subdued and embarrassed. We had only ourselves to blame, they said, for causing such a stink at headquarters the previous night. And it was our fault that they had to start out in the wee hours of this cold and miserable morning. The least we deserved, they reasoned, was a fine. "We have to produce something, you know, either a body or a ticket! Which shall it be, illegal camping or an illegal fire? We can actually give you two tickets, you know!"

So, out of the goodness of their heart we were handed the penalty of our choice, a ticket for illegal camping. To us, this charge sounded the least serious of the two. However, it came with an obligatory court appearance at a later date when the amount of the fine would be determined. Incidentally, the judge was not impressed with my lame excuse that all we had wanted to do was watch otters!

But why did we get a ticket in the first place, you might ask? At that time, it was still quite legal to camp at sites of one's own choosing anywhere in the backcountry of the park. However, unbeknownst to us, there was an obscure rule on the books which stated that random camping and open fires were permitted only beyond one mile from the nearest highway. And this is where we fell short. Our island camp was less than half that distance away from the park's main road that skirts the west side of Talbot Lake. Irrespective of the fine we got, we soon stopped going there on account of the inescapable noise of traffic.

As to otters, the lake remained a magnet for a few more years. But for them too, it came at a cost. During the late sixties, at least two otters ended up dead on the highway between Talbot and Jasper Lake. Such fatalities may well have been the main reason why the local population vanished soon after. In 1983, a comprehensive biophysical survey of Jasper Park's wildlife, sponsored by Parks Canada, listed the otter as "probably extirpated" in the lower Athabasca Valley.

Meanwhile, the highway has become many times busier than before. It is a major transportation route to the west coast for an endless procession of cars and trucks, which represent a murderous hazard to all wildlife frequenting the road corridor, day or night. It seemed doubtful that otters would ever make a comeback to Talbot Lake. Yet, they did!

After 1972, when I found the last sign, I had all but given up on the species until Brian reported fresh tracks along the Athabasca near Pocahontas. This was in the spring of 1986. Alerted by the exciting news, I began looking again in all the likely places, and in the fall of that same year I discovered a set of prints near the mouth of the Rocky River, which enters the Athabasca right by the outlet of Jasper Lake. From then on, finding otter sign became quite routine.

~

The species' come back to Jasper Park can be attributed to its compulsive urge to travel. In their ceaseless search for productive fishing holes, otters explore watercourses big and small, even tiny rivulets. During winter, these web–footed water babies are not afraid to strike out overland. Taking a childlike delight in snow and ice, they jump at any chance for a free ride down a slippery slope. Even on the level, they typically travel by alternating several running steps with a slide, coasting on their belly for a few yards. In snow, their spoor looks like a Morse code of dots and dashes. It is a singular give away. In places where otters are common and protected, the surface of frozen lakes and streams is overlain with a network of otter travel routes. It is part of an interconnected world–wide–web, so to speak, for their tribe occurs right across North America and Eurasia, with related species in the tropics.

Quite possibly, the otters that re–colonized Jasper Park arrived from the west via the Fraser River system, which forms a natural link with the Pacific coast, an otter haven beyond compare. Working their way up the turbulent headwaters of the Fraser, they reached the watershed on the other side of the continental divide via the low Yellowhead Pass, a region of gently sloping forests studded with lakes and marshes. East of the divide, the waters gather into the Athabasca, which eventually, after thousands of miles, empties into the Arctic Ocean.

One calm night, after many years of absence, a pair of newcomers to Jasper must have nosed up the outlet of Talbot to follow up a tantalizing scent. The lake is part of a wetland that abounds in whitefish, pike, and waterfowl. A chain of spring–fed ponds, like beads on a string of clear water, send their overflow down the outlet, keeping it open all winter, even in the coldest of weather, making it an otter's paradise. Although traffic fatalities still occur, the surprising thing is that most otters avoid the risk by using the culvert under the roadway. Today, Talbot is certainly not the only headquarters for otters in the park. They are reported all along the Athabasca River and several of its tributaries. For me, finding their tracks has become a relatively common event during every season of the year, but it never fails to brighten my day.

~

What is so exciting about looking at otter sign, you might ask? As hunters are fond of saying, tracks make thin soup. It all depends on what your priorities are. If you want to learn something about the hidden life of otters and understand their place in nature's scheme, seeing them is very nice but not essential. Tracks and other clues can tell the story in a subtle yet eloquent way.

One day, I discovered an otter hang–out in spring–fed Cavanach Creek that stays open all winter, at least for the first half mile or so beyond its virgin source percolating out of a crack in a mountain side. Gathering in clear pools, the shallows are festooned with aquatic vegetation, vividly green year–round. Farther downstream, depending on the severity of the season, the waters cool and freeze. Decades ago, the creek had been repeatedly dammed by beavers, but after their food supply became denuded within a safe distance from deep water, their lodges have stood vacant. Several of the old dams still hold back ponds that contain fish. Schools of brook trout, most of them no larger than six inches (15 cm) or so, have wintered in these sun–lit shallows for as long as I remember… until the otters arrived.

Working their way up from the Athabasca River, a pair of happy–go–lucky water weasels invaded the creek and struck a bonanza. They raised their brood in one of the old beaver lodges which I later found littered with the small scats of their pups. The following winter, two adults inhabited a lodge farther upstream near the best fishing holes. By spring, there were many scat piles along shore. Evidently, the family had done well. However, their prosperity had its limits.

Where I used to see schools of forty or fifty small trout, I now spot the occasional skittish minnow darting for cover as soon as my shadow falls over the water. Meanwhile, otters come and go, but my angling days are over. In the past, it was fun to catch a meal of pan–fry. I have not fished the stream for a long time. These trout do not need another enemy, and I for one do not want to compete with my friend, the otter.

I had a very similar experience in the Willow Creek district of the upper Snake Indian Valley. There, a beaver pond that used to provide a guaranteed meal ticket of a fat rainbow or bull trout suddenly stopped producing. I could cast my lure all I wanted but failed to get a single strike. At the time, my suspicion was that the secluded site had been discovered by greedy anglers. I changed my thinking after a park warden reported that he had seen three otters in the creek! The fact that these beaver ponds are so easily fished out points to the low productivity of these cold mountain waters. It also explains why the otter's propensity for travel is born of necessity.

~

Did I eventually get lucky and actually see otters, you may ask? Yes I did, quite early on in fact, during the years when Talbot Lake was still the only local otter hang–out I knew about. One of the holts Irma and I found along the island's shore indeed proved to be occupied. After landing the canoe for a brief investigation, we bent down to look into the burrow, and the invisible occupant protested with a snarl of such ferocity that we dashed back at once! As we drifted away, a Canada goose called in alarm. Perhaps it had seen the disturbed animal slip out of the holt, but we failed to notice it. Otters can dive for long distances and they have a sneaky way of staying out of sight even when surfacing, with only their eyes and nose protruding.

Our big day arrived soon afterward. Driving by the lake, we happened to spot a group of three otters drawing long v's on the water which was slick as glass. After stopping the car, we aimed the binoculars and saw the troupe clamber on top of an old beaver lodge, vanishing from view. We immediately launched the canoe and paddled silently toward the lodge, which was overgrown with grass and surrounded by water on all sides. As we floated nearer, we discovered the otters asleep, curled up together in a furry heap. Next moment, one of them gave an explosive snort and all three vanished like a brown streak, splashing into the water.

The biggest one, probably the mother, surfaced near the canoe and swam around us, hissing like a cat. At the time, I though that she was just curious. Now I know better! Otters are defensive of their young and can be aggressive toward people. One day, they attacked a swimming woman who suffered multiple bites to her legs and thighs, requiring hospitalization. Like us, the woman had spotted a family group going to rest on a beaver lodge in Talbot Lake. Failing a canoe, she had decided to strip and swim over for a closer look!

Otters are great fighters. They have to be. Slow on land, they are quickly overtaken by other predators. Turning to face their adversary, they might stand little chance against a pack of wolves, but a single wolf or coyote will think twice before tackling this snarling dynamo of tough muscle and sharp teeth. Once the otter makes its escape into water, it is untouchable.

During winter, otters that haul out with their fish onto the edge of a water hole may be accosted by thievish bald eagles or coyotes. Fascinating film footage of such attempts, much of it shot in Yellowstone National Park, has been featured on television. The pirates rarely succeeded in their objective since the otters were quick to return to water, but if well–fed, they often leave half–eaten fish on the ice that are routinely scavenged by eagles and ravens. Direct confrontations with coyotes lead to a standoff but in one spectacular film sequence, a coyote grabbed the otter by the tail just as it dived. Sinuous as a snake, it coiled back up and forced its attacker to let go again.

In Canada's North, where open water holes are few and far between, otters avoid potential enemies by spending weeks or even months under the ice. They do so especially during periods of extreme cold. According to Billy, the Indian trapper we met in Prince Albert National Park, the water level of rivers and creeks drops during the course of winter, which leaves an air space along the bank. In these dank caverns, where the blizzards never blow, the otter shares its destiny with beaver, muskrat, and mink, until they are released from their icebound prison by the coming of spring.

In maritime climates otters lead very different lives. My most rewarding observations came about during January on Vancouver Island, where I had gone to watch peregrine falcons. One morning, as I stood quietly by the shore of a pond, I became aware of some furtive activity going on beyond the rim of reeds. It turned out to be a large otter diving for food! Each time he submerged, he rolled over smoothly like a miniature Nessie, the tip of his long tail last to disappear. After every second dive or so, he came up with a small orange–bellied salamander. He went about his business with silken smoothness, leaving a minimum of ripples on the surface. The sound that had betrayed him to me was his hearty chewing and smacking!

Otters may be noisy eaters, but when hunting ducks they are deadly silent. A few days later, standing by the same pond, I happened to focus the glasses on a hooded merganser drake, arguably the most beautiful of northern waterfowl. Swimming quietly, the bird suddenly took flight and in the same instant, on the very spot, the head of an otter broke the surface. Perhaps the merganser had sensed underwater vibrations, warning it in the nick of time of the otter's torpedo–like attack.

The one place where I have yet to see an otter is my usual stomping ground, the outlet of Jasper Lake. Seeing wildlife is indeed a matter of chance and Lady Luck dispenses her favours haphazardly. Some years ago, I met a couple of young canoeists who casually mentioned that they had seen two wolves on the shore of the lake. "Oh yes, and three otters…."

This last note was added in an offhand way as if it were of little importance to them. To these young people, the adage "absence makes the heart grow fonder" clearly did not apply, but to me, as far as otters are concerned, it remains as valid as ever.

CHAPTER 22

THE KINGDOM OF BIRDS

Long ago, according to an ancient fable, all species of birds got together to choose their king. The crown would go to the one who flew the highest. Soaring to a dizzying height, the eagle surpassed all contenders. But just as he began to sail back down to earth, the minuscule wren, which had hidden itself in the eagle's feathers, sprang forth from his shoulders. Rising just a little higher on fluttering wings, the crafty dwarf burst out in cheery song and was declared king of birds.

No matter how unlikely, this fairy tale opens our eyes to the notion that the champions of the avian community are not necessarily the biggest and most rapacious. When it comes down to physical toughness during snow and cold, the facts speak for themselves. While many bird species, including the golden eagle, migrate south to avoid winter, those that stay year–round in the Canadian Rocky Mountains include tiny songsters such as the chickadee. Even on the bleakest of days, when the bark of trees cracks with frost, this jaunty bundle of feathers delights us with its cheery calls.

To better cope with the cold, the chickadee has evolved energy–saving habits. At dusk, the family retires in a tree cavity and huddles close for comfort. To further conserve heat during the night, their body temperature drops about 20° Fahrenheit (10° C). In the morning, fluffing out their feathers, the troupe of half a dozen or so sets off together. Staying in touch by means of their lilting notes, they search for dormant insects or spider eggs stuck to twigs or bark. They also eat a few seeds and, on occasion, they feast on titbits of fat gleaned from a wolf kill. The wild kingdom is intertwined indeed!

Besides seeking shelter in cavities, another survival strategy employed by the chickadee is hoarding food. Once its immediate needs are satisfied, it packs away the surplus and hides it in nooks and crannies to be saved for the worst of days when the troupe restricts its travel, again in the interest of mitigating energy losses.

Squirreling away food in times of plenty, a common practice of some mammals as well as boreal birds, is a compulsive habit of the grey jay or Canada jay. Scattering its treasure widely, it has an uncanny ability of finding them back later! Interestingly, there is a telling difference between the methods used by the grey jay and its cousin of more temperate climes, the blue jay. The latter buries its acorn or peanut in the forest litter and covers it over with a leaf or two. By contrast, grey jays avoid the ground. Instead, they make their caches in the boughs of trees or against the trunk, well above the snow line! To secure the item, the bird coats it with sticky mucous by briefly rolling it around in its mouth.

Jays start collecting winter stores early, well before the end of summer. They depend on them during the coldest of days when they, like the chickadees, restrict their movements. If a mated pair of jays has enough food in stock, they already start laying eggs in February when the woods are still deep in snow. Why so early? Young jays probably need a long adolescence to learn the tricks of the trade and gather supplies for the next season of hardship that will be upon them again all too soon.

Just as well adjusted to the cold climate and perhaps even hardier than the grey jay is the highly specialized crossbill. It may nest in any month of the year! Unlike the jay, it has no need for food storage since all its provisions are already hanging on the trees, neatly packaged in the form of cones! With its strong parrot–like beak, a crossbill can sever a cone in a thrice. Holding it underfoot, the bird inserts its curved mandibles between the scales to pry them open and extract the small seeds. If pine and spruce are hung heavily with fruit, crossbill numbers increase dramatically.

True vagabonds, they gather in flocks and go where the harvest is good. Their twittering activity and colourful plumage brighten the somber conifers.

A more sedentary recluse of the woods is the spruce grouse. Unobtrusive and spread thinly, it avoids drawing attention to itself for fear of the hawk. Its charm is its cocky confidence in the face of ground–based predators. Reacting to us as it would to a fox or lynx, it flies up to a branch just high enough to stay out of our reach, but allowing us a close look. Flexing its white–spangled tail, it beams down at us over its shoulder as if it were pleased to see us.

Like the crossbill, the grouse survives on a very monotonous winter diet: the needles of conifers which are even easier to gather than cones! Its close relative, the ruffed grouse, eats the buds of willows and poplars that are packed with nutrients. By virtue of their arboreal foraging habits neither of these "bush partridges" has to worry about snow. On the contrary, the more, the better. At night fall, they fly up for a short distance and plunge headlong into the white fluff, leaving no tracks for their enemies to follow. For extra security, they tunnel away from the entrance to a cosy bed chamber. Thus, even if the grouse is betrayed by the odour rising from the entry hole, the bird has a good chance of escaping unharmed, exploding out of its bed a few feet from the pouncing fox or lynx!

For the small and defenceless, the danger posed by birds of prey never lets up. At dusk, when the savage goshawk flies to its roost and buries its head in its feathers, the great–horned owl takes over. It is a ferocious nocturnal hunter that may even tackle the hawk if it stirs in sleep.

The variety of owl species that call the northwoods home is quite surprising, ranging from the sparrow–sized pygmy owl to the great grey. Like chickadees and woodpeckers, the smaller owls seek shelter from the weather and safety from raptors in cavities. But the largest owls tough it out in a spruce. The great grey is often about during the day. Ignoring humans, it may allow you to walk right up and admire its beautiful finery and exotic countenance. Next to thick down and bright eyes, this owl's greatest attribute is its hearing. Amplified by facial disks, its auditory powers are so acute that it is capable of pinpointing prey under deep snow. Zeroing in on the exact spot, perhaps twenty or thirty paces away from its perch, the owl pounces feet first and comes up with a vole or mouse that never knew what hit it!

On this incomplete list of winter avifauna, the raven is last but not least. A very tough customer, it can make its way from arctic tundra to fog–shrouded coasts and arid deserts. Yet, we tend to consider it synonymous with boreal wilderness. Perhaps this is so because there are so few other warm–blooded creatures to be seen. Often, during a long tramp, a raven or two is all we encounter. Fatigued and squinting in the sterile brilliance of snow and sky, our eye clicks toward the black silhouette for instant relief. The bird's strong flight and vivacious manner are reassurance that life continues despite the cold. As a scavenger, the raven benefits from the death of others. It is comrade to the wolf. Their symbiotic relationship may seem mostly in favour of the bird. But how would a lone wolf, chewing on the frozen leftovers from a hostile pack, feel without the bird's cheery attention? Like us, wolves and ravens are social creatures who need others for company and support. In the spirit world of aboriginal tribes, mankind had a pact with the raven. The bird was revered as a mystical personality, a messenger of the Gods. Its oversized image was carved on totem poles. Metaphorically speaking, I too have placed the raven on a pedestal. Without it, winter in the northwoods would be long and dull indeed.

~

Spring comes to the Jasper region in fits and starts. Driven by a blustery Chinook, pacific air suddenly floods the Athabasca Valley and may start a melt–down. Just as unpredictably, a blizzard can force a harsh retreat. By mid March, the temperature can be anywhere between 60 degrees above (15° C) or 20 degrees below (-30° C), but the first avian harbingers of the coming season are not to be discouraged.

Although Canada geese are the most visible, the bird I anticipate most eagerly is the one that would be king, the golden eagle. One calm March afternoon, I heard the hiss of what I, absent–mindedly, took to be an aeroplane passing high overhead. It was followed by the startled honking of a goose. Looking up, I saw a soaring eagle as well as a lone goose. As I followed their flight in the binoculars, two other eagles came down from the skies in sibilant stoops, both of them just missing the dodging goose!

In its rush to return north, the golden eagle, a loner by nature, meets up with others of its kind, willy–nilly forming a ragged stream of travellers along the mountainous spine of the continent. By the third week of March, it is not uncommon to see half a dozen soaring together over Roche Miette. Gaining height before sailing on, their journey may take them up the Snake Indian Valley or along the Moosehorn Range. From the Lookout Hill, I have counted as many as eighty eagles passing over during the space of one hour. On the following day only the odd one might come by or none at all. Like the arrival of spring itself, migrations are spasmodic. Depending on atmospheric conditions, the flow of eagles may stagnate or follow a different ridge route. When the wind is right, the great birds travel so high that most escape detection.

To watch this martial bird up close and in action, I had to go elsewhere, to the grassy foothills along the Bow River, just east of Banff National Park. On many a March day, I sat for hours on a high vantage point, scanning the sky and adjacent hills through binoculars. Although the golden eagle is perfectly capable of capturing a mallard in mid–air and overcoming prey as large as a fox or antelope, its most common target is the ground squirrel. This roly–poly rodent avoids winter in its own way, not by migrating but by hibernating underground. It awakens from its torpor roughly at the same time as the eagle's return. Spying on its quarry from afar, the hunting bird uses a strategy of surprise. Stooping low over the grass–covered hills, it sets its wings like a cruise missile and hugs the folds of the terrain, staying out of sight until the very last moment. Seizing the "gopher" in its claws and skidding to an abrupt halt, the hungry eagle begins to feed at once.

If well–fed, the great birds are playful and quick to harass animals too large for prey. One day, I watched spellbound as a soaring eagle planed down to just over a coyote trotting across open terrain. The dapper canid froze and stood its ground. The eagle hovered overhead, suspended on the strong Chinook. After a while, the coyote gained courage and jumped up, snapping at its opponent. The bird rose a little, just staying out of reach and descended again as soon as the coyote dropped down on all fours. Their dance was repeated several times, until the eagle lifted the siege and pulled back up over the hill.

Another eagle accosted a Hereford cow! Sailing down to the bovine that was grazing on the slopes below, the bird lowered its feet and landed on the cow's back. Panicked, the animal ran down, ridden by the eagle, wings flapping for balance. Near the base of the hill, it released its grip again and soared back up on the wind. It then selected another unsuspecting cow and rode her down too in the same spectacular and high–spirited way!

~

In the Athabasca Valley, the excitement of eagle migration is soon followed by the more unobtrusive arrival of other migrants. Hurrying northward on the heels of winter, all early birds face uncertain conditions. The first robin's song is tentative, and flocks of bluebirds may be grounded by snow squalls. Ducks sit out unseasonable setbacks on newly–formed ice. Swallows, always on the wing, might reverse direction if caught out ahead of nature's schedule.

At this latitude, the greening of deciduous trees, once started, is quickly accomplished. The buds of poplar are prepackaged in fall and coated with sticky resin to prevent drying over winter. Delicately green, the leaves unfold with the first April showers. By the time the wolf willow fills the air with its sweet aroma, the woods hum with insects. Thrushes whisper day and night. The emergence of new life, or its untimely end, takes place discreetly. The sharp–shinned hawk, flitting across our trail, is gone before we can focus our eyes.

In June, depending on the amount of rain, the mountain meadows sprout a succession of flowering plants. Long held back, their growth is rushed and may be stunted by drought. Parched grasses mature and go to seed early, yellowing in the scorching sun.

Summer is but an intermezzo. Well before the leaves drop, song birds abandon their nest sites and leave on their southward migration during the night. As winter reclaims its cold dominion, the stark forests fall silent once again, except for the cheerful notes of those few of our feathered friends, hardy enough to stay here all year.

TALES OF THE CABIN

The sturdy little hut sits well off the rough gravel road that runs to the Devona railway siding in the Athabasca Valley. From October to May, vehicle access is closed to the public, and after some of the steep, hairpin turns become choked with drifting snow, all other traffic ceases. It is then that I like going there best, hiking the ten miles (16 km) on the road or a much shorter distance through the woods, after first crossing the river.

When I was first given the key to the hut, in the interest of furthering my wildlife studies in the surrounding district, the place had not been lived in for a long time and the interior was grimy and run down. But beauty is in the eye of the beholder and I, all too pleased with the shelter, never noticed the dirt until spick–and–span Peter drew my attention to it. He inspired a gradual cleanup, and bit by bit, the looks of the interior improved although the spartan furnishings are the same today as before. The pride of place is an old–fashioned, cast–iron wood stove with white enamel sides. It functions both as cooking range and space heater although the fire box holds only a limited amount of fuel. Once it has turned to ashes, the cabin cools down until the stove is lit again in the morning. For very cold nights, a big, sheet–metal stove has been squeezed in between the range and the wood box. The upright barrel has a round lid on top and a small, adjustable air intake opening near the bottom. It can hold unsplit blocks of wood large enough to keep smouldering all night, but it looked like the stove had never been used before. One December day I decided to give it a try. By evening, a blizzard had sprung up from the north and the thermometer began to drop. In anticipation of a very cold night, I set up a pyre of fire wood in the stove's cavernous belly. It would take just one match to light, but it was not yet needed and I went to bed trying to forget my worries. I did not like being here by myself under near–arctic conditions, but there was nothing I could do now.

When I awoke by mid–night, it was plain that the cabin's interior had cooled considerably. Outside, the wind was howling in the spruces. Getting up, I shone the flashlight on the thermometer hanging by the frosted window. It had dipped to 20 degrees below

(-30° C). Shivering with cold and anxiety, I opened the lid of the stove, struck a match and lit the wax paper at the bottom of my neat pyre. I then closed the lid and went back to bed.

The stove crackled into flame like a jet plane taking off. The chimney roared and began to glow red. From my bed I could see the small, circular opening of the air–intake radiate a bright, flickering light. Its dancing reflections on the cracked linoleum of the floor had a mesmerizing effect on my drowsy mind. But just before dropping off, my guardian angel must have touched me on the shoulder for I vaguely remembered that the woodbox had seemed a little too close to the stove. Just to check, I got up and felt the side of the box, which was no more than six inches (15 cm) away from the heat source. It was far too hot to touch! I hurriedly closed the air–intake to damper the fire, picked up the water bucket, broke the layer of ice that had formed on top, and threw the contents against the box. I then unloaded most of the wood until the big container was light enough to be pushed away from the stove. It now partly blocked the door but seemed a safe distance away from the heat.

Next morning, I was shocked to see that the side of the box was scorched black and the varnish blistered. The wood must have been near the point of spontaneous combustion. Surely, if I had not got up in time, the entire cabin might have gone up in flames! And even if I had escaped the inferno and survived the blizzard outside, I would have been the laughing stock of the warden service for having burned the cabin down in my very first winter!

Henceforth, I learned to adjust the barrel stove just right. Staying warm required a minimum of work since all the fire wood I needed was supplied, courtesy of the wardens. All I had to do was split the blocks with a big axe, a task that President Ronald Reagan thought was the most relaxing hobby in the world.

~

Comparatively minor but persistent problems were presented by mice and flies. The latter, as the saying goes, came out of the woodwork soon after the place heated up. Catching them, one by one, was a routine chore, especially during March. Mice were never absent for very long. At first, they had the run of the place, some of them so bold as to dart up and down my bare back while I was on the verge of sleep! In the still of night, their rattling in nooks and corners could be annoying. Moreover, they messed up the food supplies and, in my absence, gnawed holes in the mattresses. Regretfully, I had to resort to a mean, old stand–by, the never–improved–upon snap–trap baited with peanut butter. Unfortunately, one of the cute innocents became caught by the tail and dragged itself under the stove where it was hard to reach. Another one died of unknown causes and was not found until it began to decompose and create a nauseous stench. Dealing with mice became more of a worry after a Health Canada warning that the deer mouse was a carrier of the Hanta virus, which could be transmitted to humans by contact with the rodents or their droppings. The disease is hard to diagnose and has proved fatal for a small number of campers and farmers across Alberta.

Officially, in a national park, all killing of wildlife is against the law, and so is feeding the animals. This made me a double sinner since I ended up giving the trapped mice to the foxes! Scrounging around human habitation is as normal for a wild fox as scavenging from wolf kills. Sooner or later, the dead mice that I disposed of by the wood pile disappeared. Coming and going by night, all the fox left were his tracks. In my desire to set eyes on my welcome but secretive visitor, I devised a way of making him knock on the door! After placing the mouse on a log in view of the window, I attached it with a thin rope to a piece of firewood hanging from the cabin's wall. Any animal making off with the bait was bound to pull the string and lift the stick of wood. Upon release, it would automatically swing back against the wall!

My crafty knocker worked like a charm. At some point during the night, the rattling of the stick against the cabin made me jump out of bed at once. Shining a flashlight through the window I saw nothing at first, but presently, the sly fox sneaked up to the frozen mouse lying in the snow and made off with it. This time, he either severed the string with his sharp teeth or freed the bait by pulling it out of the loop.

It was not long until Reynard showed up each night and occasionally during the day. Seeing him trot up the path to the cabin on a sub–zero morning, his fur and whiskers freshly dusted with snow, is among my most treasured memories! One winter, my main visitor was a cross fox featuring a dark belly and throat. There were also black foxes in the area with grizzled guard hairs. These were the fabled silver foxes of the fur trading era, their skins worth their weight in gold. Interbreeding between blacks and reds produces a mix of offspring, making it possible to recognize individuals by their pelage. One winter, I saw at least three different foxes in the yard.

In order not to disappoint my customers in mouse–free times, I enticed them with a piece of chicken or other meat. It was fun while it lasted but it ended rather abruptly, not on their account, but by the arrival of the competition. This proved to be another common but elusive local animal, the pine marten. It became so bold and quick to respond that the foxes never had another chance. As soon as the local marten heard the cabin door or smelled the smoke from the chimney, he came up the path. Fearless, he grabbed the bait without hesitation and refused to let go even if I stood in the doorway and pulled on my end of the string. Growling, his jaws clamped on the meat, he followed me right into the cabin. This is where I drew the line, so to speak, and stopped feeding the animals altogether. Ever since then, the occasional fox or marten comes by and takes the dead mouse if there happens to be one, but I have resolved not to interfere in their lives. In the same spirit, I feed neither the squirrels nor the jays and chickadees.

~

Although I usually stay at the cabin alone and love its solitude, the company of others has its own rewards. It creates a comfortable sense of security, and animated conversation helps pass the time. There have been, however, less positive moments. One day, Brian caused me great anxiety when he failed to show up as expected. He was supposed to arrive after a long detour via Pocahontas. At dusk, I forced myself to stop worrying about the possibility that he might have come to grief in the cold and deep snow. We had no way of contacting each other. Perhaps he had changed his plans and gone back to the highway. Just before nightfall, a casual glance outside revealed his dark form coming up the path. Never in my life have I been so happy to see another human!

In turn, I caused similar worry to Peter after I tracked a moose and stayed out until dusk. However, instead of being happy at my safe return, he scolded me severely, no doubt with good reason, for my lack of consideration. In his mind he had already been to the police to report a missing naturalist.

On the bright side, another companion once made my day when I broke the filling of a tooth. The cut was so sharp that my tongue began to bleed and I was unable to eat or speak.

"File it down," Alan said and produced his Swiss Army Knife. Like a hurt child, I opened my mouth, expecting him to do the job, but he insisted that I do it myself. It was the right advice. Applying the file to the exact spot, the problem was solved very quickly, saving my skin as well as the day!

If alone, the winter evenings could be very long. After a day outside in the cold, all I felt like doing was staring into the light of a candle and contemplate. Even reading took more energy than I could muster for more than half an hour or so. Thoreau's *Walden*, one chapter at the time, was appropriate for the occasion, but the stories of Andy Russell were easier to digest. A paperback edition of his *Trails of a Wilderness Wanderer*, small enough to fit in my pack, included an anecdote about an old–time trapper in the western mountains where snow lies deep. Spending the cold night outside, he built a huge fire and after the embers had sank well down into the snow, he placed long pine poles over the hole as a platform for his sleeping robe. He probably ended up smoked like a side of pork!

Such stories of hardship made me all the more happy with the warm cabin and I savoured the privilege of being here in the wilds of the Canadian Rockies, in a National Park where all species of wildlife were free to come and go! Although we now take these paradise–like conditions for granted, wildlife protection is a product of relatively recent times. Things were very different before the park was established!

~

The known history of the area goes back to 1810 when David Thompson, following his Indian guides, became the first European fur trader to set foot in the Athabasca Valley. The ruins of historical Jasper House, built by Thompson's successors, were only a few miles away from the cabin. To find out about the wildlife inventory of that era, I obtained the diaries of one of the earliest postmasters of Jasper House. Micro–film copies of his notes were lent to me by the archives of the Hudson's Bay Company, which is still in business, but now on a very much wider scale. The handwritten entries were easy to read since the spelling of the English language has changed little since then. From 1827 to 1831, Jasper House was manned by a Scot named Michael Klyne. According to one of his contemporaries, Klyne was a jolly old fellow with a native wife and a large family. His post was termed "a miserable concern of rough logs." As many as eighteen people slept in one of the two rooms, probably like sardines in a tin. However, it was not uncommon for the Jasper House people to go to bed hungry.

Postmaster Klyne often experienced great difficulty feeding his family, servants and unexpected visitors. In his account of the fall and winter of 1830–1831, he mentions that his full–time metis hunter saw only three elk, of which he shot two. Additional proof that hoofed mammals were hard to find in those days can be inferred from the fact that the hunter frequently returned from his wide–ranging forays without having seen a single animal or even a fresh track! Klyne's diaries also contain graphic references to starving Indians. During the course of the winter, he was visited by several bands of Stoneys or Assiniboines, a tribe that had recently immigrated here from the Lake of the Woods region in Ontario. Some arrived with beaver skins for trade, others without, but the Indians were invariably very hungry. Klyne helped where he could although his store of proviand was often low.

By early March, three family groups of Shuswaps or Snake Indians showed up at the post. They were the area's original inhabitants, still living the way they had done for generations, using primitive weapons and stone tools. "They are starving all the time," Klyne wrote.

On March 25, he penned: "A Shuswap arrived almost dead of starving. I gave him a little meat to eat for himself and a little to take to his family. I can not give him much. I have little myself. In the evening, my hunter arrived. Saw nothing."

On April 25, a Shuswap woman showed up with her three children. "A few days past, two other children and her husband died of hunger…."

Two weeks later, during the night, the Shuswap woman left, abandoning two of her kids at Jasper House. "I sent after her to come for her children but no–one could find her track."

The Indians were not necessarily helping each other, on the contrary. Much like wolf packs that compete for territory, the immigrant tribes were hostile toward the aboriginals of the area. Conflict was all the more serious since the Shuswaps or Snakes were only armed with bows and spears, whereas the newcomers had obtained rifles from the white traders in the east. Under the pretext of making peace, the Stoneys invited the able–bodied men of the Snakes to their camp near the mouth of the river that now carries their name. Once they were seated around the fire, they were shot. The Stoneys then rushed over to the Snake camp intent on murdering the remaining members of the band. Only a few managed to escape and three young women were taken prisoner. Later, at a trading post far downstream the Athabasca, a HBC employee helped the three captives escape, practically without clothes or supplies. Two of the women were never heard from again, but one survived for two years in the wild, feeding herself with plant roots, berries, and rabbits she managed to snare. The woman was eventually discovered by another Indian. Subdued after a furious fight, she was led away to Jasper House. A few years later, she was reunited with some survivors of her Snake Indian tribe who happened to come by the post. This drama unfolded in 1830 just a few miles away from the Devona cabin. No doubt, many other human tragedies that took place here remain forever lost in the so–called good old days.

~

A slow improvement in the living conditions of people and animals began in 1907, after the establishment of Jasper National Park. Wildlife populations that had suffered from the relentless hunting pressure of native and European trappers, were protected and al-lowed to recover. Elk, imported from Yellowstone National Park, were reintroduced, and as the herds increased, wolves made a comeback on their own. They too have now found their rightful place, and I had been their witness and defender.

During the recent past, after my arrival in the park, subtle changes have continued to unfold, not only in the fauna but also in the flora. As noted in chapter seven, the Willow Creek meadows have become choked with bushes and trees. A similar plant succession from open ground to forest is underway in the lower Athabasca Valley. Spruce seedlings are getting established on open sites. Everywhere, trees are maturing and growing taller, a slow but inexorable process that is beginning to have a serious impact on the view from the Lookout Hill. Formerly, I could overlook a large expanse of meadows and river flats, which made it possible to watch the passage of a wolf from one side to the other. Now, I might catch no more than a brief glimpse here or there between the spruces that are blocking much of the view. As to the elk census, I used to be reasonably certain of seeing all of the members of a herd. Today, only part of the group is visible between the trees at any one time.

Nature is forever evolving, and the activities of humans continue to play a part in the process of change, for better or for worse. In the past, natives and whites kept forest fires going from year to year. As noted by David Thompson, the environs were in a constant state of conflagration. For that reason, as shown in photographs of Jasper House taken in 1872, the mountain sides of Roche Ronde were largely bare. Now, they are green with conifers. A return to fire as a management tool to restore the grasslands remains a controversial and little tried option.

As seen from a greater perspective, on a geological time scale, the Rocky Mountains are a very young landscape. Less than ten thousand years ago, during the last ice age, the valleys were scoured out by glaciers. Only the highest peaks towered above the chaos. Over the millennia of change, the ice sheet melted and created rivers such as the Athabasca. As long as its life–giving waters flow, change will be followed by more change.

It is my hope that one condition will remain the same and everlasting: the continued protection and respect for all living things. To the credit of our Johnny–come–lately race, these values have become part of the changes affecting Jasper National Park, much to the benefit of the animals and their watchers alike.

Relevant Publications

Ballantyne, E. E. 1955. Rabies control programme in Alberta. Canadian Journal of Comparative Medicine 20:21–30.

Banfield, A. W. F. 1974. The mammals of Canada. National Museum of Natural Sciences, Ottawa. 438 pp.

Burpee, L. J. (Ed.). 1907. The Journal of Anthony Hendry, 1754–55 – York Factory to Blackfeet Country. Royal Society of Canada. Pages 307–354.

Carbyn, L. N. 1975. Wolf predation and behavioural interactions with elk and other ungulates in an area of high prey diversity. Ph. D. Thesis. University of Toronto. 233 pp.

Carbyn, L. N. 1989. Coyote attacks on children in western North America. Wildlife Society Bulletin 17:444–446.

Cowan, I. M. 1947. The timber wolf in the Rocky Mountain National Parks of Canada. Canadian Journal of Research 25:139–174.

Dekker, D. 1983. Denning and foraging habits of red foxes and their interaction with coyotes in central Alberta. Canadian Field–Naturalist 97:303–306.

Dekker, D. 1985. Hunting behaviour of golden eagles migrating in southwestern Alberta. Canadian Field–Naturalist 99:383–385.

Dekker, D. 1985. Responses of wolves to simulated howling on a homesite during fall and winter in Jasper National Park, Alberta. Canadian Field–Naturalist 99:90–93.

Dekker, D. 1985. Elk population fluctuations and their probable causes in the Snake Indian Valley of Jasper National Park: 1970–1985. Alberta Naturalist 15(2):49–54.

Dekker, D. 1985. Wild Hunters. Canadian Wolf Defenders Publication. 224 pp.

Dekker, D. 1986. Coyote preys on two bighorn lambs in Jasper National Park, Alberta. Canadian Field–Naturalist 100:272–273.

Dekker, D. 1986. Wolf numbers and colour phases in Jasper National Park, Alberta: 1965–1984. Canadian Field–Naturalist 100:550–553.

Dekker, D. 1987. The not–so–natural history of Jasper National Park. Park News 23(4):26–29.

Dekker, D. 1989. Otters return to Jasper National Park. Alberta Naturalist 19(4):141–142.

Dekker, D. 1989. Population fluctuations and spatial relationships among wolves, coyotes and red foxes in Jasper National Park, Alberta. Canadian Field–Naturalist 103:261–264.

Dekker, D. 1997. Wolves of the Rocky Mountains – From Jasper to Yellowstone. Hancock House Publishers, Surrey, B.C. 208 pp.

Dekker, D. 1998. Pack size and colour morphs of one wolf pack in Jasper National Park, Alberta, 1979–1998. Canadian Field–Naturalist 112:709-710.

Dekker, D. 2001. Two decades of wildlife investigations at Devona, Jasper National Park, 1981–2001. Unpublished report in collaboration with Wes Bradford and other park wardens. 52 pages plus appendix.

Dekker, D., W. Bradford, and J. Gunson. 1995. Elk and wolves in Jasper National Park, Alberta – From historical times to 1992. Pages 85–94 in: Ecology and conservation of wolves in a changing world. Editors: L. N. Carbyn, S. H. Fritts and D. R. Seip. Canadian Circumpolar Institute, Edmonton, Alberta. Occasional Publication No. 35. 642 pp.

Geist, V. 1971. Mountain sheep – A study in behavior and evolution. The University of Chicago Press. 383 pp.

Glover, R. (Ed) 1962. David Thompson's Narrative, 1784–1812. The Champlain Society, Toronto. 410 pp.

Gunson, J. R. 1992. Historical and present management of wolves in Alberta. Wildlife Society Bulletin 20:330–339.

Henry, J.D. 1986. Red fox – The catlike canine. Smithsonian Institution Press, Washington, D.C. 174 pp.

Herrero, S. 1985. Bear attacks: their causes and avoidance. Nick Lyons Books, Winchester Press, USA.

Mech, L. D. 1967. The Wolf. The Natural History Press, New York. 384 pp.

Mech, L. D. and M. Korb. 1978. An unusually long pursuit of a deer by a wolf. Journal of Mammalogy 59:860–861.

Messier, F., and M. Crete. 1985. Moose–wolf dynamics and the natural regulation of moose populations. Oecologia (Berlin) 65:503–512.

Messier, F., C. Barrete, and J. Huot. 1986. Coyote predation on a white–tailed deer population in southern Quebec. Canadian Journal of Zoology 64:1134–1136.

Murie, O. J. 1954. A field guide to animal tracks. The Peterson Field Guide Series. Houghton Mifflin Company, Boston. 374 pp.

Peterson, R. O. 1977. Wolf ecology and prey relationships on Isle Royale. U.S. National Park Service Science Monograph Series 11. 210 pp.

Russell, A. 1970. Trails of a wilderness wanderer. Ballantine Books, New York. 241 pp.

Stelfox, J. G. 1971. Bighorn sheep in the Canadian Rockies; a history 1800–1970. Canadian Field–Naturalist 85:101–122.